The Compleat Loo

An underground public convenience in suburban London, *circa* 1900 – with cast-iron railings and ventilation lamps.

The Compleat Loo

A Lavatorial Miscellany

Roger Kilroy

BARNES
& NOBLE
BOOKS
NEW YORK

This edition published by Barnes and Noble Inc, by arrangement with Greenwich Editions, Bibliophile House, 10 Blenheim Court, Brewery Road, London N7 9NT.

First published in the UK in 1984 by Victor Gollancz Ltd.

ISBN 0 7607 0416 3

Printed in Singapore

M 10 9 8 7 6 5 4 3 2 1

Library of Congress in publication data available upon request.

1996 Barnes & Noble Books

Acknowledgements

The cartoons on pages 8, 12, 24, 42, 48, 56, 62, 70, 72, 80, 88 and 94 are by Larry. In addition, the author and publishers would like to thank the following people for permission to include in this book illustrations which are their copyright.

Armitage Shanks Ltd, page 35 (top and bottom right); BBC Hulton Picture Library, pages 16, 26, 27 and 28; The Belgian National Tourist Office, page 79 (bottom left); Bibliothèque Nationale, Paris, front of jacket; Tim Bishop, page 67; The Bodleian Library, Oxford, page 10 (left); Harold K. Bowen/Gladstone Pottery Museum, Stoke-on-Trent, pages 23, 32 (left), 33 (left), 37 and 38 (top and bottom left); British Aerospace, page 85; Richard Bryant, page 40; Clos-o-mat (Great Britain) Ltd, page 96 (bottom left); Dominic Cooper/The National Trust, page 18 (top); Dent & Hellyer, page 18 (bottom left); Department of the Environment, pages 11 (top) and 19 (right); Douglas Dickins, page 11 (bottom); The Committee of the Egypt Exploration Society, page 10 (bottom right); Mary Evans Picture Library, pages 2, 64 (right) and 79 (top left); The Fotomas Index, pages 17, 30 (top), 32 (right) and 46 (bottom); Anthony Green Ltd, page 90 (left); André Goulancourt, page 90 (right); M. Hoviz, page 79 (bottom centre); Hygolet UK Ltd, page 96 (right); Ideal Standard Ltd, pages 6 and 41; Lucy Irvine, page 77; The John Johnson Collection, Bodleian Library, Oxford, pages 50 and 92; Keystone Press Agency, page 79 (right); Lucinda Lambton Library, pages 18 (bottom right), 21, 29 (right), 31 (left), 35 (top left), 39 (left and top right), 46 (top), 51 (left), 64 (bottom left), 65, 66 (top and bottom left) and 84 (top right); Lucinda Lambton Library/The National Trust, page 39 (bottom right); Leicestershire Museums, page 47 (bottom); Macdonald & Jane's, page 51 (right); The Mansell Collection, pages 29 (left) and 68; The National Trust, page 45 (top); North of England Open Air Museum, Beamish, Co Durham, pages 52 and 53 (left); Putnam & Co, page 93 (left); Royal Bathrooms, page 36; Royal Doulton (UK) Ltd, pages 33 (right) and 34; The Royal Navy Submarine Museum, page 84 (top left and bottom); Science Museum Library, page 20; Alec Scott/Dunsdale Lodge Antiques, page 47 (top); Sea Containers Ltd, page 83 and back of jacket; Secker & Warburg Ltd, page 93 (right); Severn Valley Railway, page 82; Sheridan Photo-Library, pages 10 (top right) and 19 (left); Swedal Leisure (UK) Ltd, page 53 (right); Transbyn Ltd, page 96 (top left); Twyfords Bathrooms, pages 31 (right) and 66 (bottom right); Reece Winston, page 64 (top left); Wolferton Station Museum, page 38 (right). The photograph of the close stool at Hampton Court Palace on page 45 is reproduced by gracious permission of Her Majesty the Queen.

Contents

Introduction

It was doing the research for the graffiti books that first aroused my interest in loos. I never actually clocked up the number of hours I spent in public lavatories, but there must have been a good few, and I am glad to say I managed to avoid both arrest and infection. I became fascinated by the range of facilities available – from ancient channel urinals requiring delicacy of aim to massive marbled thrones which always produced a comfortable feeling of relaxation. Seated one day in peaceful contemplation I began to wonder what life must have been like before there were flush lavatories. What did people *do*? Thus began my research. I rushed off to the library the very next day, and started reading books about loos.

I had vaguely thought that flush lavatories were invented in Victorian times, and was astonished to discover that they dated back to 2000 BC. Then, apart from a few medieval monasteries which had WCs of a kind, nothing happened until the sixteenth century when Sir John Harington invented his – and it still took 200 years to catch on. Still, eventually it did, thank goodness, and now, looking to the future, there are loos that chop up waste, loos that can reduce their content to a tenth of its bulk, even loos that do away with the need for loo paper. One wonders what will come next. And if you've always been puzzled about what astronauts do when they need to go to the loo, look no further, the answer is in this book. I hope you may find a quiet and peaceful place in which to read it.

R.K.

ET TU BRUTE
CENTURIONS
SYRUP OF FIGS
SYRUP OF FIGS

Early Times

We do not know when man first started using special places in which to relieve himself but stone huts dating from Neolithic times which have been unearthed at Skara Brae in the Orkneys have what appear to be crude drains leading from recesses in the walls that have led people to believe that they were latrines. And at Knossos, in Crete, in the palace of King Minos, a splendid bathroom was constructed about 2000 BC with cleverly designed plumbing and latrines. One of them even appears to have had a wooden seat and an earthenware pan like a modern lavatory, with a reservoir for flushing water. It took England nearly 3,500 years to come up with anything as good.

The ancient cities of the Indus valley around 2,500 to 1,500 BC also appear to have had houses with bathrooms and water-flushed lavatories which were connected to a sewerage system. The city of Akhenaten at Tel-el-Amarna in Egypt, which dates from about 1350 BC, had latrines with seats made of limestone and keyhole-shaped orifices, under which were removable containers to catch the waste.

The Romans also used this shape in the stone seats of some of their latrines, as can be seen from the photographs. The Romans, of course, have always been noted for their bathing habits, and at the height of the Roman Empire used something like 300 gallons of water per head of the populace per day. Present-day Londoners use about 50 gallons per head per day, but this probably means that the Romans were more wasteful with their water than we are. The Romans constructed latrines for both public places and private houses, and some of the latter were water-flushed. In AD 315 there were 144 public latrines in Rome. Some of them may have been vessels which could be carried away and emptied, as in those days urine was collected and used to help remove the grease from clothes, and people left vessels in public places to help the collecting process.

Rome also provided sanitation for its armies, an excellent example of which can be seen at Housesteads Fort on Hadrian's Wall. The main latrine here could seat twenty men side by side. It measures 30 by 16 feet internally, and has a continuous wooden trough running along two sides over which it is believed wooden seats were fixed, probably in the form of a continuous wooden bench.

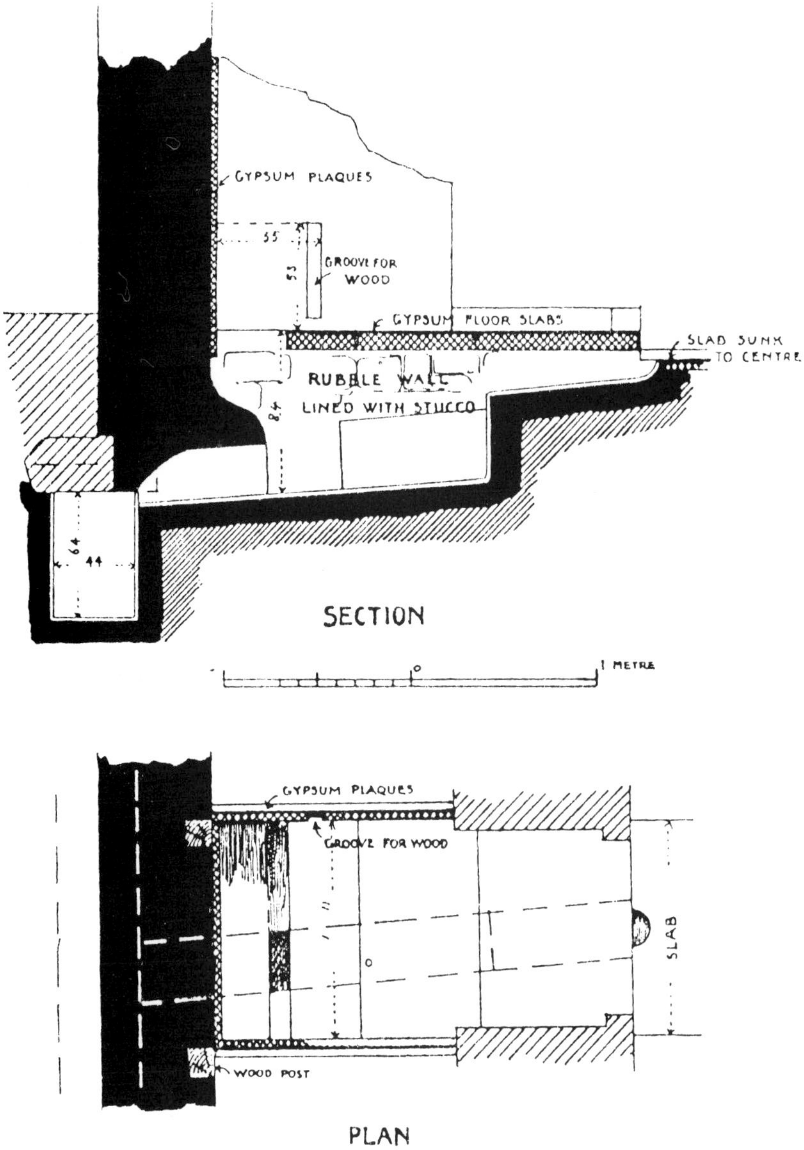

Plan and section of latrine in the Palace of King Minos, Knossos, Crete, which dates back to about 2000 BC.

Public latrine at Ostia Antica, Italy, first century AD.

Latrine at the city of Akhenaten, Tel-el-Amarna. The shaped limestone seat looks a little more comfortable than its Roman equivalent.

The Roman latrine at Housesteads Fort on Hadrian's Wall. At the feet of the users was a water channel into which they dipped the sponge sticks they used instead of toilet paper, and both the trough and the water channel were supplied with rainwater to flush them by means of an elaborate system of stone water tanks and pipes.

Another Roman lavatory at Leptis Magna, Libya.

PLYMOUTH BOWLING CLUB
W.C
THE ARMADA CAN WAIT
SPANISH ARMADA

The Dirty Years

After the fall of Rome it was said that 'for a thousand years Europe went unwashed', as the ancient civilisations were swept away by the invading uncivilised hordes from the north. In country areas people retired 'a bow's shot away', and in general relieved themselves outside. In later years, when they built privies, these were situated in small outhouses, away from the house, and the waste material went into a cesspit beneath, where the liquid filtered away into the earth and the solid matter was removed from time to time and used on the fields. This kind of system prevailed until recent times in many rural areas, and indeed in England today there are places where the contents of the cesspool, when pumped out by the council workmen with their tanker, are spread on the land.

Town living, however, was a different matter. Here waste products of all kinds were discharged into the streets and watercourses. There were both public and private latrines of a sort, including a public one on London Bridge which discharged straight into the river, and gave rise to the saying that the bridge was built for wise men to go over and fools to go under. Some dwellings did have privies that discharged into pits, for housing ordinances of 1189 required that the pits, if not walled, should be situated at least 5 ft 6 in from the boundary between the houses; if walled, they were to be at least 2 ft 6 in away. But in the twelfth century these must have been the exception rather than the rule, for it is recorded that the Fleet and Walbrook rivers in London became extremely offensive and impassable, and it was not until after the Black Death epidemic in the fourteenth century, which wiped out between a third and a half of the total population of England, that people stopped discharging all the sewage directly into the rivers and began to build cesspits as a matter of course. The cesspits had to be cleaned out periodically by men called gongfermors, who seem to have been well paid and have taken a pride in their trade. A Thomas Brightfield who lived in London in 1449 built a kind of water closet, but the idea does not seem to have caught on.

An illustration of what life was like in those days is provided by a quote from Boorde – The Medieval Privy 1400–1500:

> 'Beware of draughty privys and of pyssynge in draughts,
> and permyt no common pyssyng place about the house –

> and let the common house of easement to be over *some water* or else elongated from the house.
>
> Beware of emptying pysse pottes and pyssing in chymnes.'

By 'chymnes' was meant the fireplaces. And Leonardo da Vinci produced a proposal for ten new towns, in which the staircases in the buildings were to be spiral to discourage people from soiling the landings.

If people didn't use a privy or public latrine they used chamber pots or 'close stools', the contents of which were thrown out of the windows at night with the words 'Gardy loo' (*Gardez l'eau*). For this reason it was considered unwise to be out too late at night – the returning reveller might get his just deserts!

The cleanest places in medieval times were probably the monasteries. Many had quite sophisticated piped water systems, and sensible rules about washing before eating. Plans still exist for the water supply and drainage system of Christchurch Monastery at Canterbury, which was completed about the middle of the twelfth century, and was so designed that the waste water from the wash places and the rainwater from the roof was directed through a main sewer than ran under the latrines and flushed away the waste. Latrines in monasteries were called 'necessary houses' and often occupied quite large areas of the building. The one at Canterbury was 145 ft long. Often the sanitary wing was linked to the main building by a bridge, providing a separation which was sensible on health grounds but must have proved rather inconvenient. Since the monks had to follow a rigid timetable each day the necessary house had to allow for a large number to use it at the same time, and in some places seats were set back to back in a double row. But more usually they were built in a single row, with partitions between them, and a small window for ventilation. An account of the monastery at Durham, written in 1662, speaks of 'as many Seats on either side as there were little Windows in the Wall'.

Castles, too, had latrines. Called 'garderobes' they were usually built either within the walls (which were several feet thick) or jutting out from the walls, with an opening leading straight to the moat or ground below. If there was no moat the garderobes often emptied into a pit far below, which had to be cleaned out from time to time. This direct connection to the outside world made the garderobe a cold and draughty place and to make matters worse it often had a stone seat. Small wonder that Sir John Harington (see below) wrote of: 'A godly father sitting on a draught, To do as need and nature hath us taught'. But it caused problems other than draughts, for it was a break in the castle's defence system, and several castles were captured by people climbing up the garderobe shaft, including the Chateau Gaillard, which was designed by Richard I.

The word 'garderobe' is similar to our word 'wardrobe', and meant the same thing. It is interesting to note that even in the Middle Ages people were coy about their lavatorial expressions. To speak of the garderobe then was rather like asking the whereabouts of the cloakroom now.

In 1596 the man who might be called the 'father of the water closet', Sir John Harington, a godson of Queen Elizabeth I, published a book called *The Metamorphosis of Ajax*, which explained 'how unsavoury places may be made sweet, noisome places made wholesome, filthy places made cleanly'. 'Ajax' was a pun on 'a jakes', i.e. a latrine, and the book described a valve water closet he had invented. It consisted of a cistern made of stone or brick, depicted with fish swimming in it; a pipe leading from the cistern to the underneath and rear of the seat and controlled by a stop cock; a vessel 2 ft deep, 1 ft broad and 1 ft 4 in long, like a large chamber pot, made of brick, stone or lead; and a waste pipe with a lockable sluice gate leading out of it to a drain. The sluice gate could be locked to prevent children playing with it. The whole thing cost 30s 8d – in decimal coinage about £1.60. Harington specified that if water was plentiful it should be used often, if scarce, then once a day would be enough, even if twenty people used it, and 'your worst privy may be as sweet as your best chamber'. Queen Elizabeth had one installed in her palace at Richmond, and a copy of the book was kept chained to the wall by the side of the WC.

No one seems to know why the idea did not catch on, but it was to be another 200 years before the WC was reinvented and came into general use.

By the eighteenth century some of the great houses had water closets, though these were few in number and often situated outside. In France they were known as *cabinets d'aisance à l'anglaise* or *lieux anglaises*. The closets of those days were often made of metal, but around the turn of the century they began to be made of glazed pottery, and as their use became more widespread the most splendid designs and ornamentation appeared. The age of Victorian sanitary magnificence had almost arrived.

Old London Bridge, *circa* 1680 from an engraving by J. Kips.

A medieval street scene. It was unwise to venture out too late at night!

Fountains Abbey, Yorkshire. The arches show the position of the latrines. Twelfth century.

Marble latrines at Agra, India, in the fifteenth century. Perfumed water ran along the channel, and vases containing perfumed water were placed on the two octagonal bases for hand washing.

Garderobes at Chipchase Castle, fourteenth century.

Garderobe in the Castle of King Rene, Tarascon, fourteenth century.

The garderobe chute on the east face of the White Tower, Tower of London.

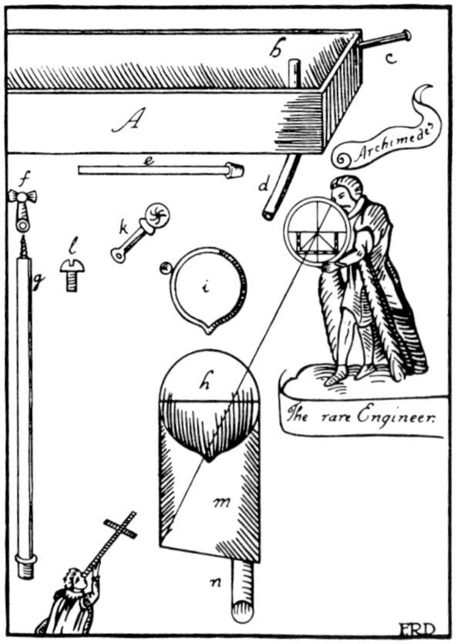

Drawings showing Sir John Harington's 'Ajax'. A is the cistern, D the seat, H the pot, L the sluice, M and N the vault into which it falls.

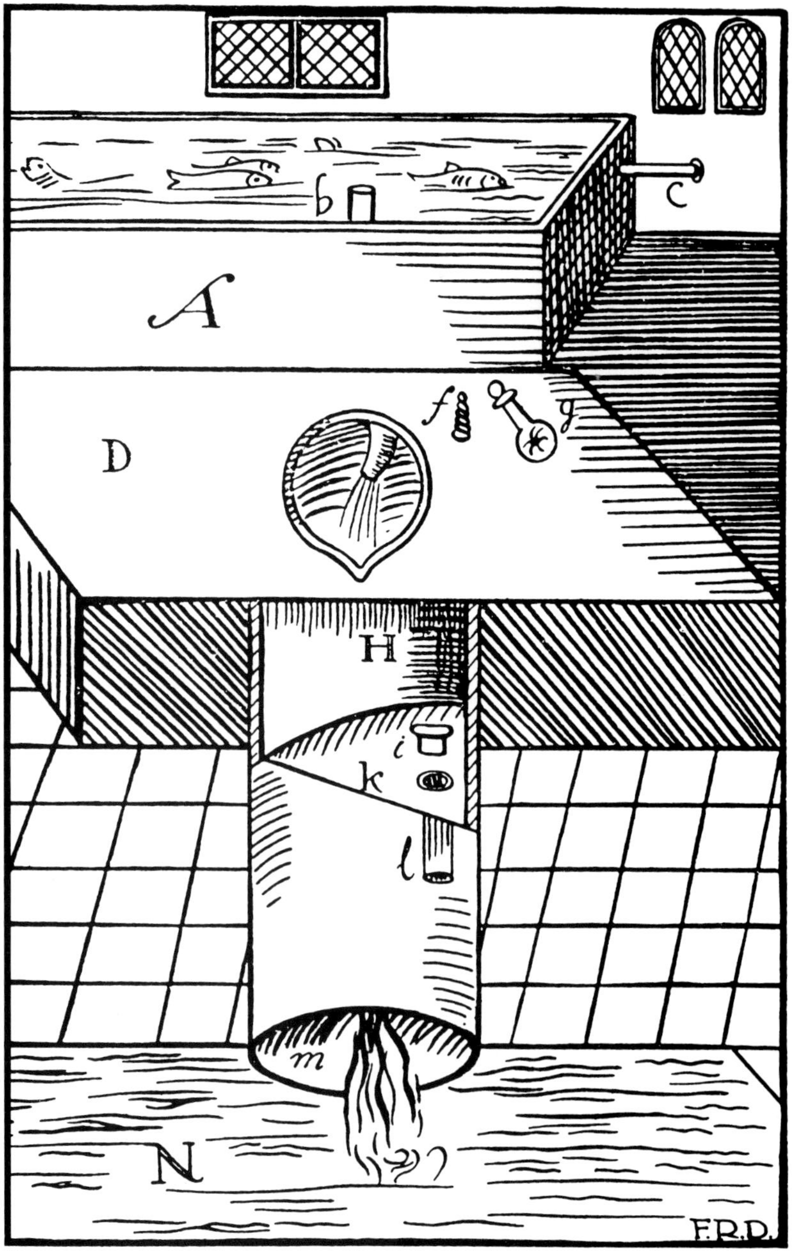

A six-seater privy at Chilthorne Dormer, Somerset, which dates from approximately the late-seventeenth to mid-eighteenth century.

A one-and-a-half-seater privy for a mother and child in a churchyard at Selling, Kent.

The pan closet was one of the earliest of the eighteenth-century WCs. The upper pan, which contained a few inches of water, fitted into a lower container and by pulling a handle was supposed to discharge its contents into the lower container and out into a trap below. But the lower container often retained the waste, and the closets fed sewer gas into the house.

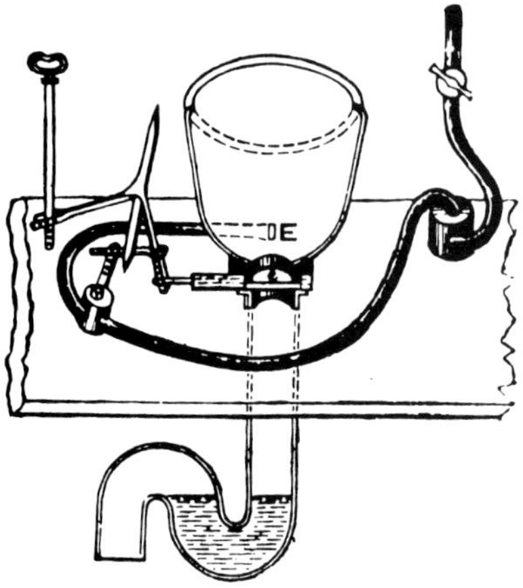

In 1775 Alexander Cummings patented a valve closet which had an overhead water cistern and a valve which connected with the flush mechanism to open the bowl's outlet into a syphon trap. The idea was that every time the WC was used the water in the trap and its contents were flushed away, but in practice this was not always the case.

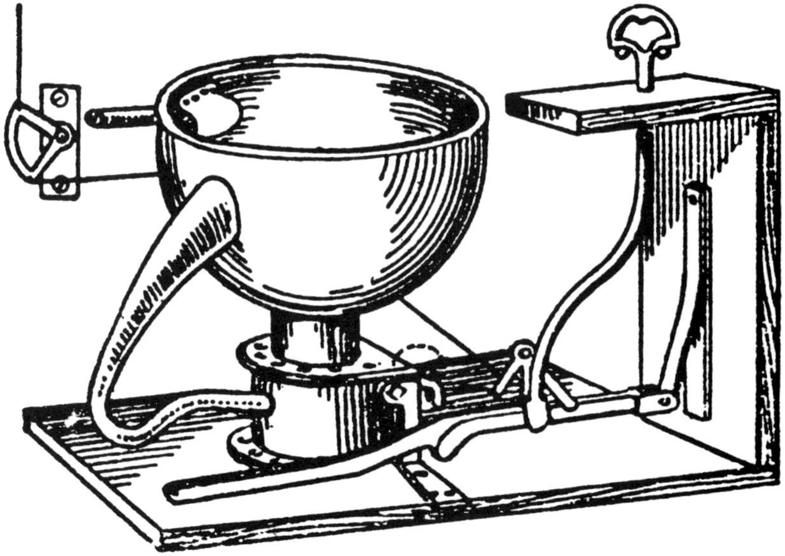

An improved version of Cummings's closet was patented by Joseph Bramah in 1778. Bramah's closets continued in production until 1890.

A wooden-cased valve closet of the late eighteenth century.

THREE BRONTE SISTERS
LOCKED IN A LAVATORY

Into the Modern Age

The nineteenth century saw many developments in sanitary engineering. The rise in population of the big cities, and the acres of inadequate housing which resulted, gave rise to many health problems. Rows of terraced houses, often of the notorious 'back to back' type, were served by one water pump and one privy for several families. The privies discharged into cesspits which were emptied by 'night men', and if the cesspit was situated under the house, as was often the case, its contents may have had to be carried through the house to be emptied. Not surprisingly there were many outbreaks of disease, notably the cholera epidemics around the middle of the century, which claimed 30,000 victims in London alone. This dreadful state of affairs focused public attention on to the insanitary state of Britain's towns and gradually the cesspits were replaced by proper sewers.

The early WCs had many faults, and it wasn't until the 1870s and 1880s that efficient ones were produced. It was in the 1870s that a Mr Twyford of Hanley came up with the idea of the 'washout' closet, which had a shallow bowl which retained a little water and a flush (which was said to lose most of its force in emptying the bowl and therefore didn't take everything that it should with it). This problem did not affect its sales, and it was very popular. Stevens Hellyer, an author as well as a plumber, produced an improved version of the Bramah valve closet in the same year. This was called the Optimus, and its working parts, which looked extremely complicated, were hidden beneath specially built mahogany cabinets, or even specially built chairs. The problem of noisily flushing WCs was improved by a contemporary, J. R. Mann, who produced in the same year a syphonic closet which had a fast flush followed by a slower one, and was considerably quieter in operation than its predecessors.

This was followed in 1884 by George Jennings's 'Pedestal Vase' which won a Gold Medal Award for being 'as perfect a sanitary closet as can be made'. In a test its two-gallon flush carried cleanly away ten apples averaging 1¼ inches in diameter, one flat sponge 4½ inches in diameter, plumber's 'smudge' coated over the pan, and four pieces of paper adhering closely to the soiled surface. A modern plumber would have a fit if he thought all that were being put down the loo.

The following year Twyford produced the washout Unitas closet, and that and Jennings's Pedestal Vase were the first to be fixed in position without being covered by a wooden casing, which meant they could be cleaned more easily and any faults could be detected and repaired.

The washdown closet, the one still used today, was claimed to have been invented by D. T. Bostel in 1889. Its efficient flush carried away the waste without impediment.

One problem that vexed the water authorities of those days was the fact that the WCs used a great deal of water. The flush pull operated a valve to release the water, and not only did the valves often fit badly, but they could be left open to allow a continuous flow. And so stepped into history the wonderfully named Thomas Crapper, mistakenly believed by many people to have invented the WC. What he did do, however, was to perfect a 'Valveless Water Waste Preventer' – the kind of mechanism to be found in modern WC cisterns which allows water to flush the WC only when required.

But the most noticeable feature of these late nineteenth-century WCs was the splendour of their bowls and cisterns, a few of which can be seen on the following pages.

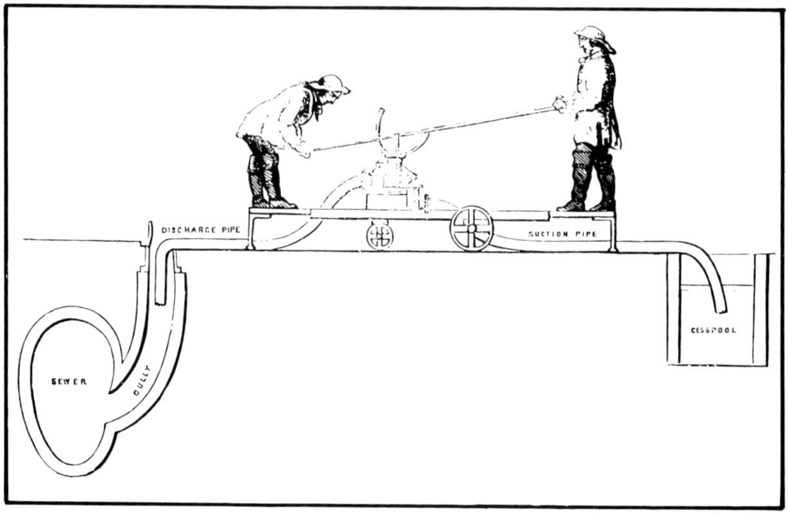

A diagram showing how cesspools were pumped out, London 1861.

Two London 'night men' of the 1840s.

'Flushing the sewers', London, 1840s.

The underground reservoir at the Metropolitan Main Drainage Works at Crossness, opened by the Prince of Wales (later Edward VII) in 1865.

The 'Pottery Sanatory Water Closet' made by John Ridgway & Co in 1851. It was made of hard vitreous pottery, had a syphon trap to keep it free from smell, and was said to cost only a third of the price of the common closet.

A three-seater privy in the garden of Kelmscott Manor, Oxfordshire, reputed to have been used by William Morris and Dante Gabriel Rossetti.

Hellyer's 'Optimus' valve closet.

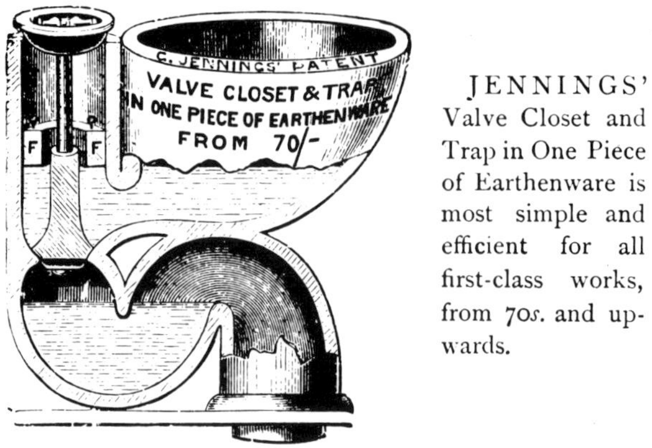

An advertisement for Jennings's valve closet.

The Country Seat

by Christopher Curtis

They were built like emporia in the reign of Victoria
In the castle, or manor, or grange,
With their seats made of wood, which have gamely withstood
Pressures greater than mere winds of change.

Boys with bats, balls, or oars, sportsmen sporting 12-bores
Gaze in rows from the walls on the sitter.
Draughts and damp old stone tiles mean today's stately piles
Suffer badly from cold that is bitter.

But, forget all the strain, pull the gleaming brass chain
(With a porcelain handle, no less),
And, released by a piston, from within a vast cistern
Comes a roar – and you're flushed with success.

Twyford's 'Unitas' of 1885.

One of the most beautiful Victorian loos, Twyford's 'Unitas' in 'raised oak' design, with a pattern of blue flowers and leaves inside. This design was produced in 1888.

The 'Combination' by Doulton Bathrooms, 1889.

A contemporary drawing captioned 'Water-closet, with "Sanitas" Enclosure, isolated from the walls and floor.'

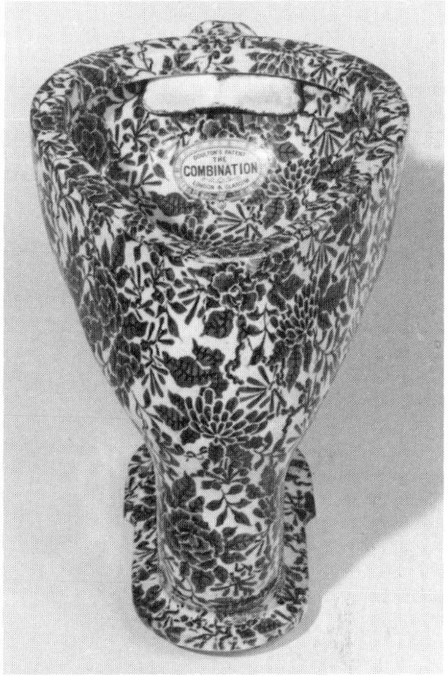

Doulton's 'Combination', produced in 1898 and used for the next sixty years. It features a blue magnolia design and cost £5 7s 3d.

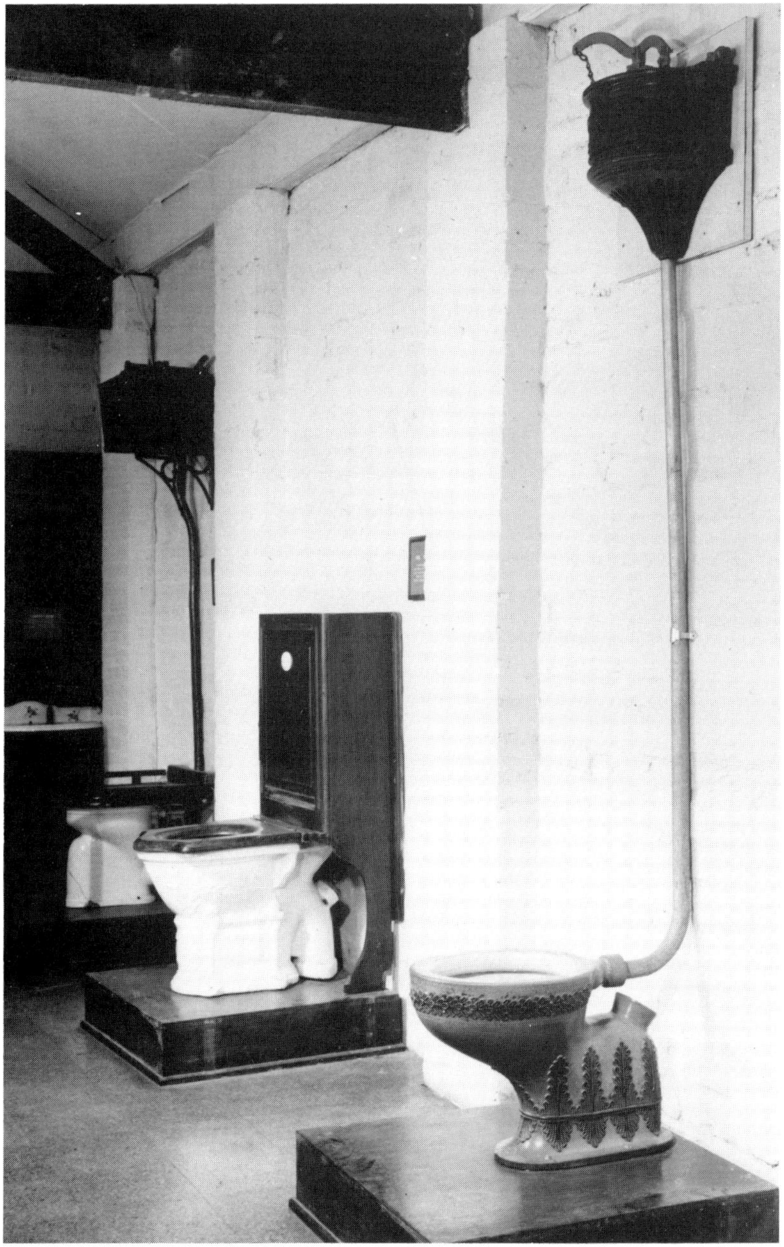

The 'Paisley' washdown closet of salt-glazed stoneware with a cast-iron cistern. Made by Doulton & Co of Lambeth, *circa* 1890.

In the centre is the 'Citizen' of 1895; in the background the 'Sanitas' of 1908.

Doulton's 'Simplicitas', also made in 1898 and with blue decoration.

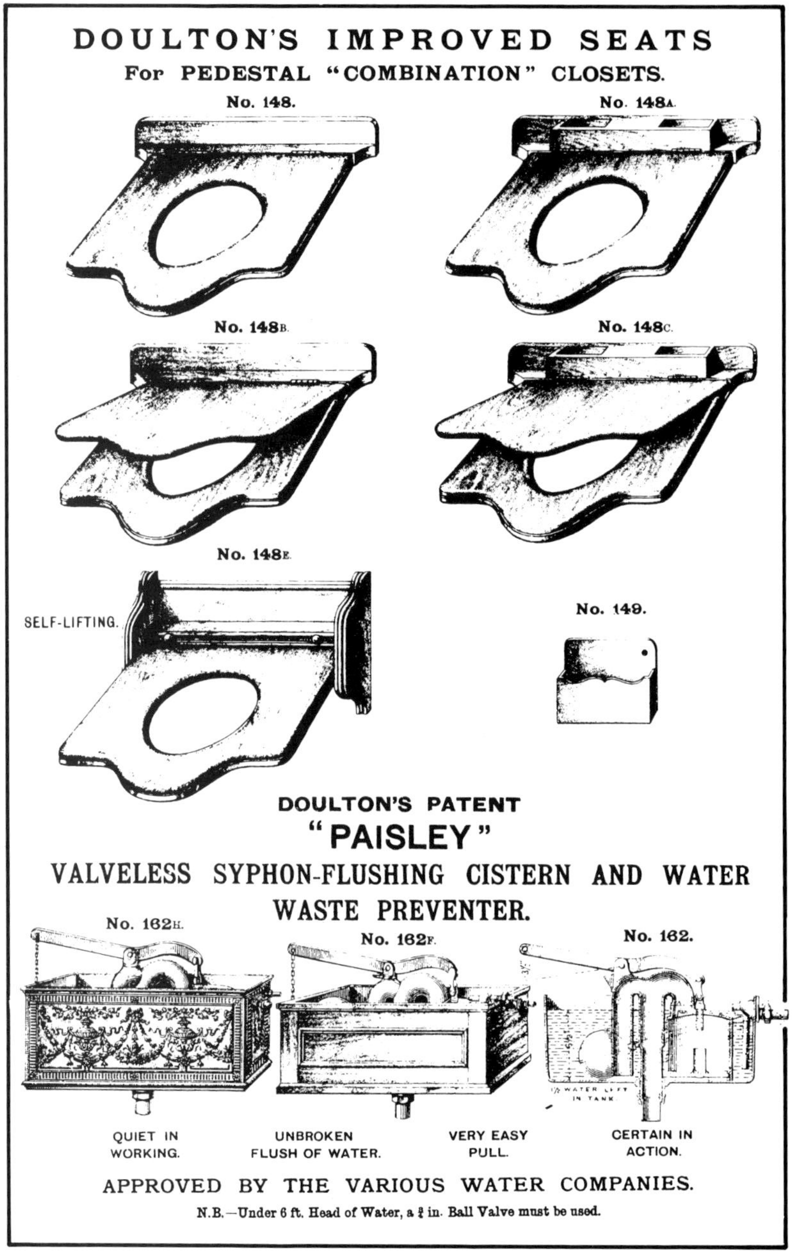

Seats and cisterns with the 'water waste preventer' from Doulton's 1898 catalogue.

Shanks's 'Mulberry Chrysanthemum' 'Torrens' cistern of 1899, described as being of 'high sanitary efficiency and beautiful appearance'. This example is at Keir, Scotland.

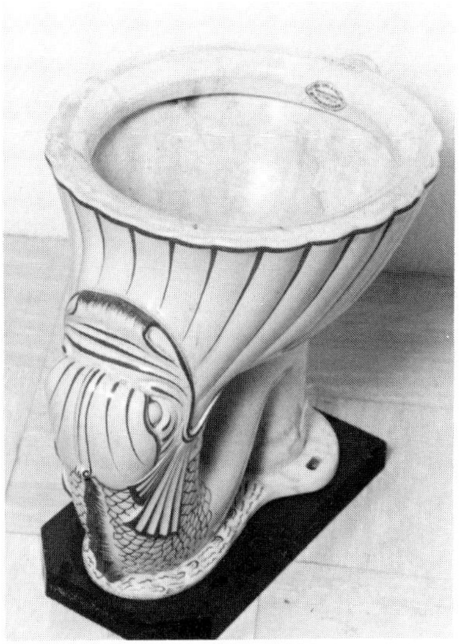

The magnificent 'Dolphin' by Stock Sons and Taylor of Birmingham.

A Shanks's WC in a blue floral design on white.

Two elegant WCs of the 1890s produced by Royal Bathrooms.

An asset to any bathroom in the 1890s – the 'Latestas' suite with washdown closet.

George Jennings's 'Closet of the Century', 1900.

The beautiful 'Oracle' washdown closet made around 1900. The manufacturer is not known.

King Edward VII's urinal made by George Jennings and bearing the 'flags of all nations'. It is at Wolferton Railway Station, on the Sandringham Estate.

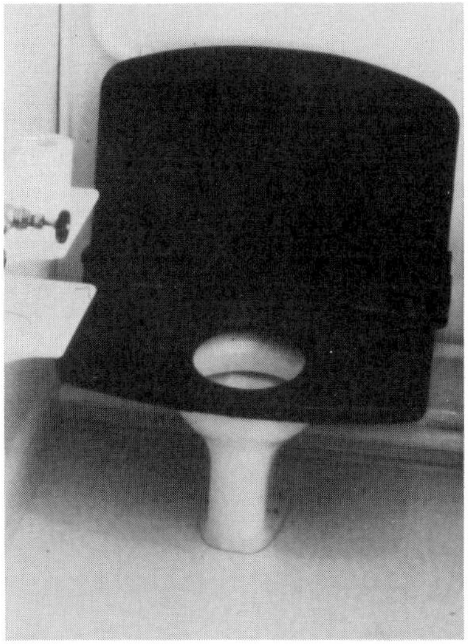

A magnificent mahogany seat at Lady Astor's house, Rest Harrow, Kent.

A lavatory casing by Lutyens over a valve closet in Castle Drogo, Devon, 1918.

A hand-painted lavatory seat from Oxfordshire, 1970s.

A wall-hung WC in a modern bathroom.

The WC seen here next to the space-age shower is made of fibreglass.

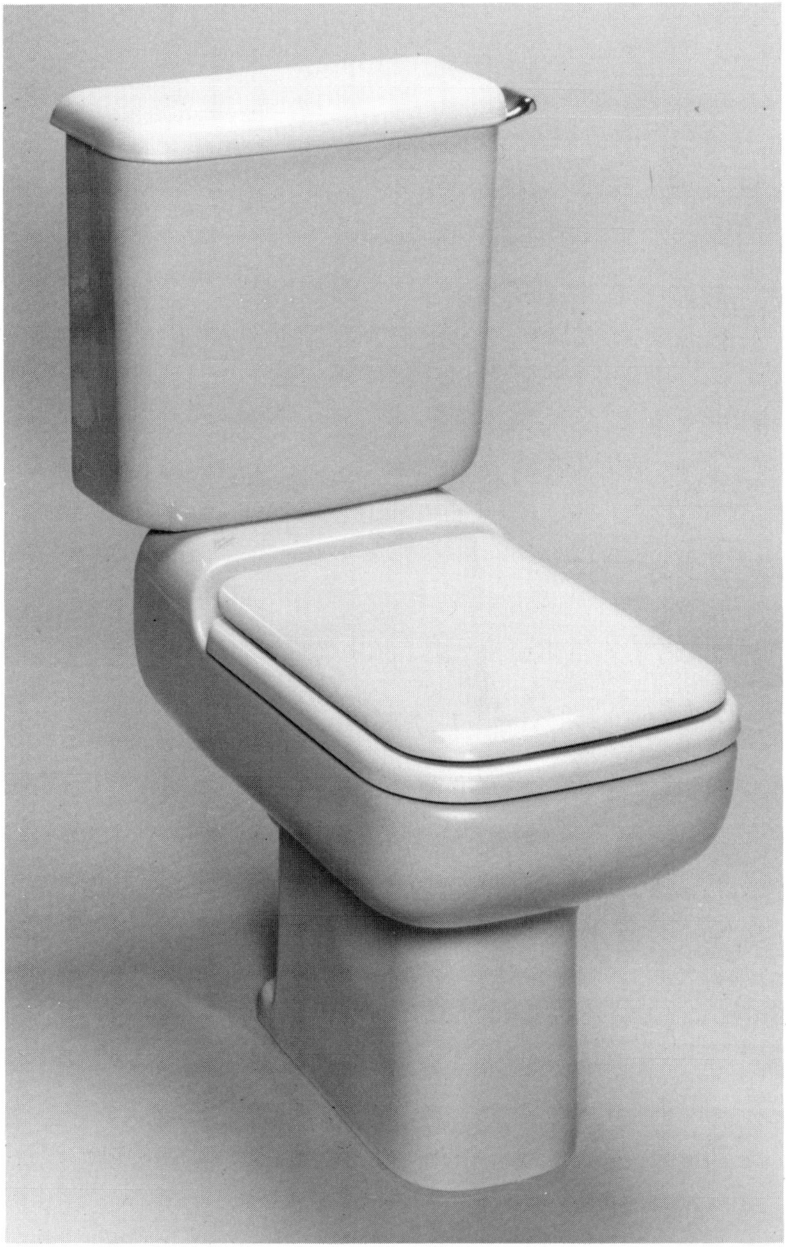

OLIVER TWIST ASKS FOR MORE

'An Indispensable Item'

Why does your brother hide under the bed?
Because he thinks he's a little potty.

This is a familiar children's joke, for the chamber pot under the bed was an indispensable item of furniture in most houses in the days of cold outdoor lavatories. In the north of England the pot was called a 'goes under', or, in more refined households, an 'article', but it is more commonly called a pot, potty or jerry.

Chamber pots are said to have been invented by the Sybarites because they were too lazy to go outside to the privy. They must have been very attached to them, as they are said to have taken their own personal pots with them to parties.

In medieval times the chamber pot seems to have been openly displayed in the bedroom, with no attempt made to conceal it, but in those days pots were also installed in the dining room, possibly concealed in a cupboard in the sideboard, or behind the window shutter recesses, for the use of men who stayed late round the table drinking. Cabinet makers such as Hepplewhite made special cupboards in which pots could be concealed – little square cupboards on legs, that must have been instantly recognisable.

The designs for chamber pots are many and varied, as are the materials from which they were made. Usually they were of earthenware or metal – probably tin or pewter – but some were made of silver. In a recent sale at Sotheby's, an eighteenth-century silver pot which had belonged to the Earl of Warrington (who owned several engraved with his arms), was sold for £10,450. By the early nineteenth century the pots had become elaborate in their design. They were often decorated with patterns of flowers, and some carried the raised design of a frog inside. Others were decorated with portraits of notables such as Napoleon or Gladstone, and a favourite, and rather daring pattern, was that of a large eye, or a face, and the following rhyme:

> Use me well, and keep me clean,
> And I'll not tell what I have seen.

Carriages had their own pots hidden under the seats, and sometimes the seats had holes in them to give direct access to the pot, hole and pot being concealed by the seat cushions.

As well as chamber pots, people have, from ancient times, used urinals. In the Middle Ages these were sometimes made of glass, and doctors would examine their patients' urine through the glass to determine their state of health. Jugs were sometimes used, as has been proved by the deposits of urine salts found in them. Ladies used bourdaloues, little receptacles shaped rather like gravy boats, in the eighteenth century, and were said to carry them in their muffs. One wonders what they did with them after use.

A grander and rather less portable form of the chamber pot (though kings and nobleman did transport theirs from place to place) was the close stool. This looked like an elegantly upholstered box-shaped chest, with a hinged lid which opened to reveal the pot. Close stools were often upholstered in velvet and had padded seats, and the hinged lids locked to prevent anyone other than the rightful owners from using them.

The close stool gradually gave way to the pot hidden inside a piece of furniture. Its presence might be heavily disguised – the French at the time of the Dutch wars produced a model which looked like a pile of books with the title *Voyages au Pays Bas* volumes I to IV. Another French collection of 'reading matter' was entitled *Mystères de Paris*. These hiding places for chamber pots came to be called commodes, and were usually in the form of a chest of drawers, cupboard or chair, though, according to lines written by Jonathan Swift, hiding the pot from view proved in vain if it betrayed its presence by its smell!

Two close stools from Dunham Massey, dating from the early eighteenth century. The one on the left is made of mahogany; that on the right of walnut.

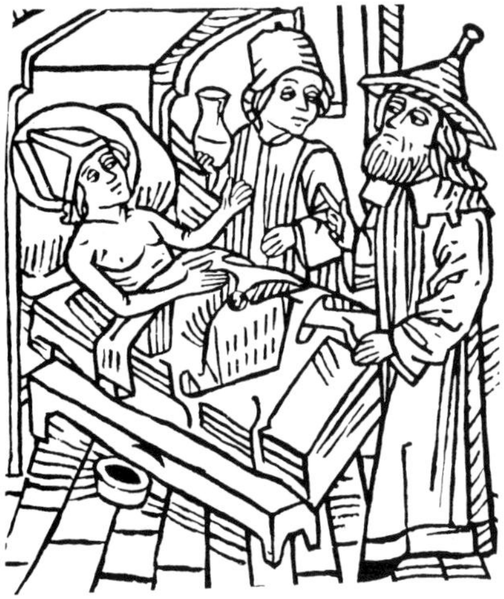

A medieval bedroom scene. The chamber pot is under the bed, and a glass urinal containing urine is being held for the doctor's inspection.

A close stool covered with red velvet from Hampton Court Palace.

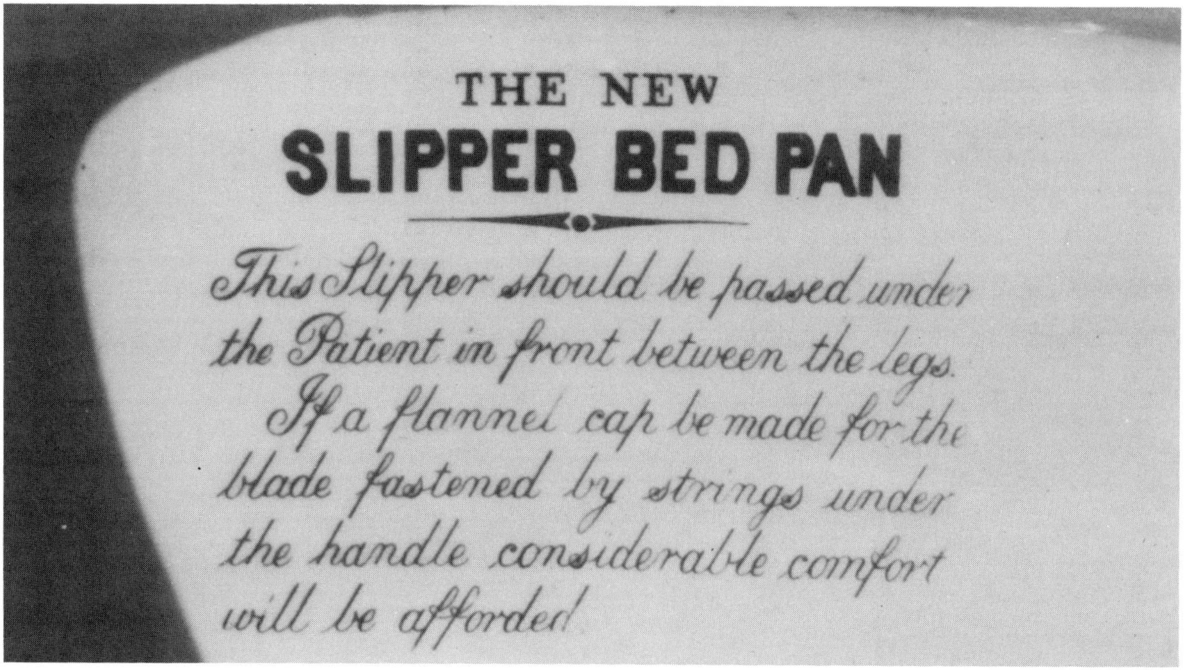

Lord Londonderry's slipper bed pan. Nineteenth century, from Wynard Park, County Durham.

A French commode disguised as a pile of books on a stool.

A very rare Victorian chamber pot, *circa* 1865–70, finished in copper lustre with applied flower decoration. It is 6 in high, 10½ in wide, and was made in north-east England, possibly in Sunderland.

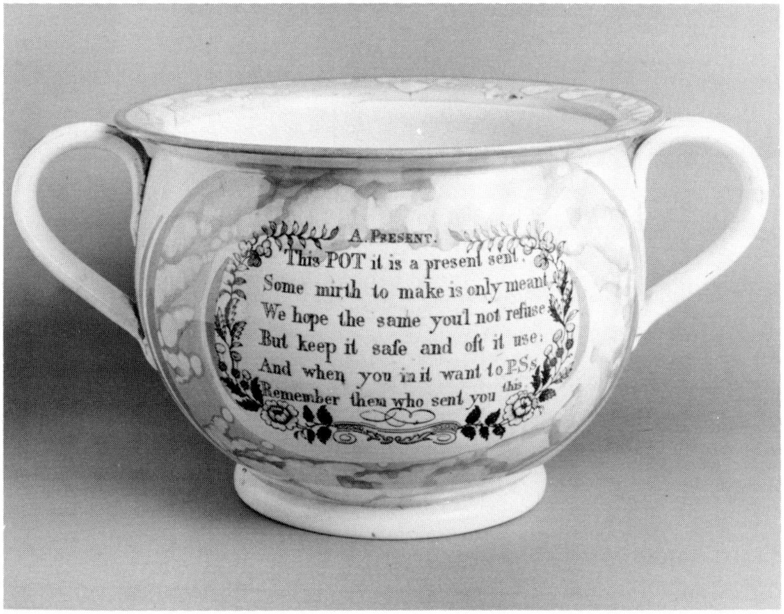

A Sunderland lustreware chamber pot.

GHIA
FLUSH
SUN
LOO
PAPER

Alternative Technology

At the same time as Messrs Twyford, Hellyer and Jennings were busy perfecting their water closets other inventors occupied themselves by devising alternatives. In 1860 the Reverend Henry Moule invented the earth closet. This consisted of a wooden seat fitted over a pail, with behind it a hopper filled with fine dry earth, charcoal or ashes. Pulling a handle released a quantity of the earth to cover the contents of the pail which would be emptied every so often, dried out and used as fertiliser. Some variations worked automatically, so that the earth was released by the movement of the seat as the user stood up, but it was thought that this might not be a good idea in the sick room, where earth closets seem to have been considered especially useful, as the sudden noise made by the falling earth might upset the patient! Indeed, a W. Liddiard invented a commode with a 'Patent Self-acting Foot-board' to operate the earth mechanism, to avoid 'the unpleasantness felt in a sinking seat'. The same inventor had his wits about him when it came to schoolchildren, too, and produced an earth closet which could be operated from a distance for use in schools, to prevent the children from playing with the closets until the earth hoppers were empty. A dry-ash commode of 1873 used the ashes collected from the firegrates, which sounds like a good idea but must have been extremely dusty.

The Centre for Alternative Technology in Wales publishes leaflets on the alternatives to water in sewage disposal, both because it is a waste of water and a potentially valuable fertiliser is lost. There are compost toilets which depend on time and heat to evaporate the moisture and destroy the pathogens in the sewage; biological toilets which use enzymes to dissolve the waste which is then disposed of in a drain field; bio-electric toilets which use heat and circulating air to reduce the waste to one-tenth of its original volume; oil-flushed toilets which use recyclable mineral oil; incinerating toilets which reduce sewage to sterile ash; variations on the traditional cesspit and septic tank; earth closets as mentioned above; and a 'pee can' to collect urine with which to fertilise the garden. But anyone tempted to rush off and recycle their own waste should take care – composting takes at least a year and the material requires careful handling if it is not to cause disease.

The Reverend Henry Moule's earth closet of 1860.

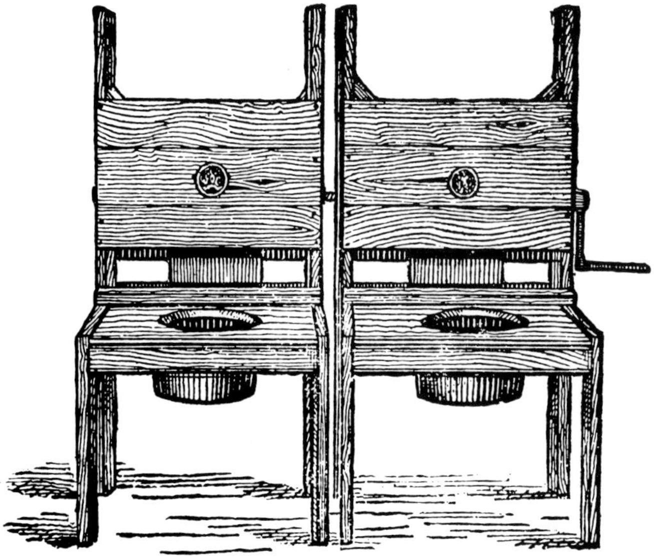

W. Liddiard's double earth closet for schools. A rod ran through the centre of each closet – any number of which could be fastened together – and was attached to a handle which worked the earth mechanism. A wire, which could be run to any part of the school, could be fitted instead of a handle to stop children playing with the mechanism. Working the apparatus eight or nine times a day was said to be 'sufficient to avoid unpleasantness'.

An earth closet at Knole, Kent, which was still used in the 1970s. A hopper empties sand into the buckets.

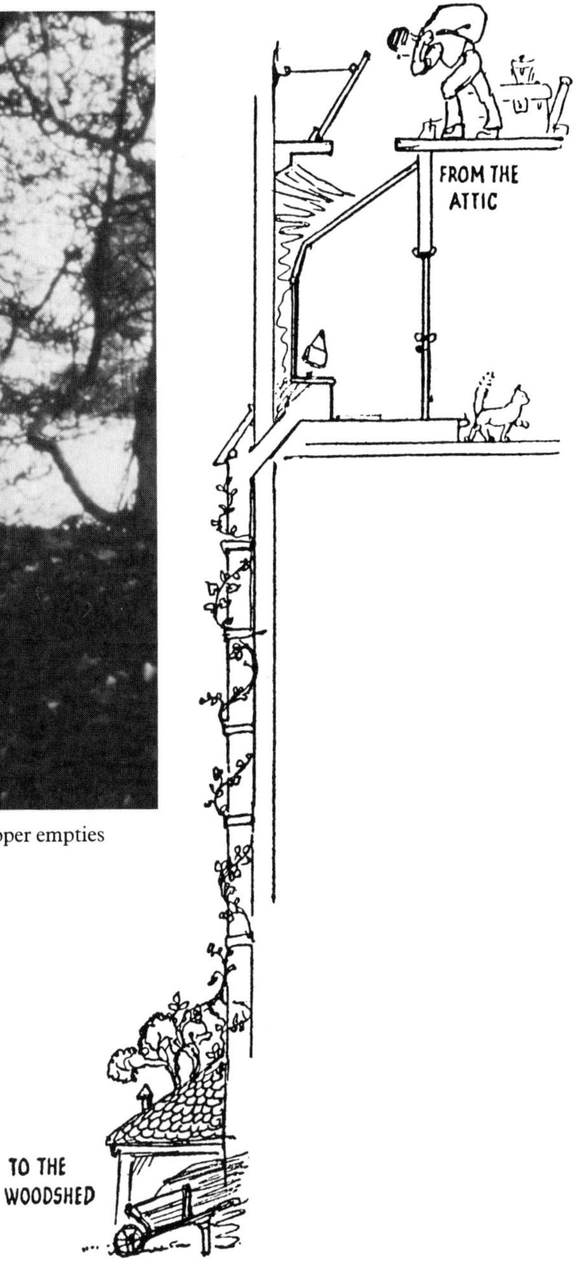

A diagram of an Oxfordshire earth closet of the 1930s. Dried earth was carried up to the attic and tipped into a hopper below a trap door in the floor. The closet on the next floor down was placed against the wall of the house, and the outlet went all the way down to a compost heap in the wood shed. The design was very similar to that of the garderobes of medieval castles.

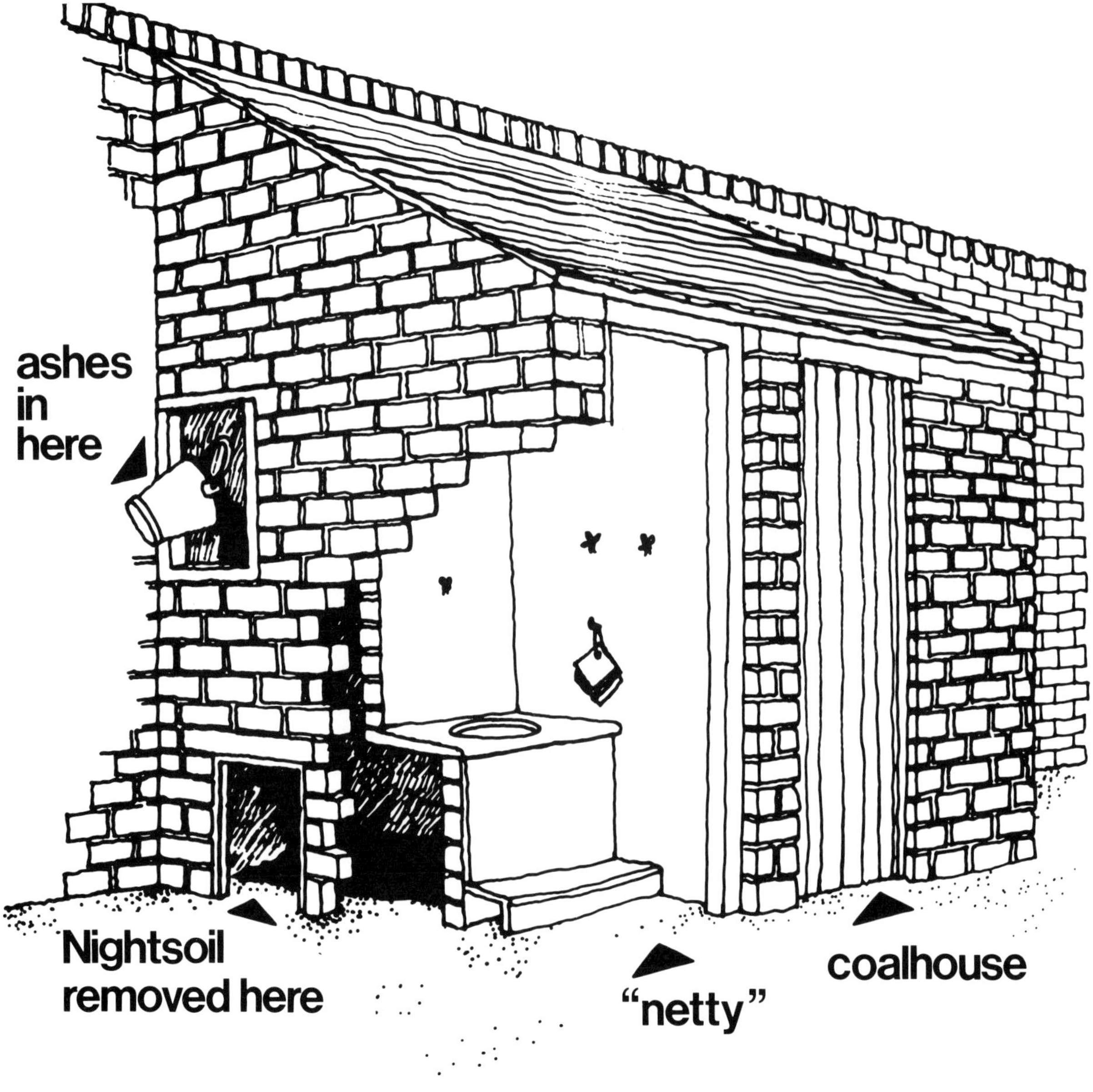

A diagram of a Geordie earth closet or 'netty', in the back garden of a miner's house in County Durham.

The Merus 'Lectrolav, originally produced in Scandinavia for use in holiday cabins and now marketed in the UK. It reduces the waste to a tenth of its original volume, evaporates urine via a charcoal filter to avoid smell, and, if used in a weekend or holiday home needs emptying only twice a year. After composting the waste can be added to the soil.

A row of netties and coal houses, County Durham.

A Regular Life

The current vogue for high-fibre diets brings home the fact that many people are obsessed with their bowels. Before bran there were horrors such as liquid paraffin, senna pods and castor oil, and anyone who has ever spent any time in hospital will not need reminding that the question the nurses never fail to ask every single day is, 'Have you *been*?' (Nor the relief felt if you can answer 'yes' and thus avoid the danger of being given a suppository!)

This obsession is nothing new. Excavations at St Albans Abbey in 1924 revealed a pit 25 ft below the cloister floor in which were found buckthorn seeds, a powerful aperient, proving that even medieval monks worried about constipation.

In times past people believed that emptying the body of waste products was the key to good health, and they took purgatives to help them achieve it. A Dr Brandreth, in a book published in New York in 1871, explained the theory thus:

> The body wears. Movement causes waste. The blood carries new material to repair the waste, and it reloads itself with worn-out parts which it discharges through the appropriate vents. The worn-out parts must be expelled from the body daily, or the blood will become impure.

Because of this, said Dr Brandreth, disease can be removed by purgation, or 'cleansing', and he invented special pills for this purpose. He claimed they would cure all kinds of ailments, from jaundice to concussion, and reproduced in his book glowing testimonials from those cured. Dr Brandreth became a very rich man!

BRANDRETH'S PILLS

USE FOR THE TEMPORARY RELIEF

of CONSTIPATION

Brandreth's Pills have been in use since 1759, when they were first compounded in England by Dr. Brandreth. Many people have difficulty in obtaining a laxative which will agree with them and at the same time produce the desired results. We recommend Brandreth's Pills as an effective laxative.

●

EACH PILL CONTAINS

0.0547 grams Aloes
0.0212 grams Gum Gualac
0.0071 grams Gum Arabic
0.0044 grams White Castile Soap
0.0026 grams Capsicum
0.0212 grams Cascarin
0.0132 grams Colocynth
0.0052 grams Extract of Sarasparilla

Each Box Contains 36 Pills

DOSE FOR ADULTS

As a laxative take one, two or three pills at night when retiring. As a purgative more may be taken under a doctor's direction. Use for children as your doctor prescribes.

Manufactured for
Allcock Mfg. Co. Ossining, N.Y.

●

WARNING

Laxatives should not be used in cases of nausea, vomiting or acute abdominal pain as these may be symptoms of appendicitis. Laxatives taken continually form a habit.

Brandreth's Pills – contents and dosage.

An advertisement for Brandreth's Pills which appeared in a little book of puzzles published in New York in 1900. The answers to the puzzles were not given – they could only be obtained by sending a wrapper from Brandreth's Pills or other patent medicine produced by the company.

"Its no good
ye standing
on this seat
the crabs in here
jump 16 feet"
WC

Seats of Learning

As a compiler of six books of graffiti and someone with a lifelong interest in the subject I felt that some examples absolutely *had* to appear in this book, for, as we all know, the lavatory is the place that inspires most of it.

Graffiti seems to be as old as man's ability to write. If you visit the ruined cities of Pompeii and Herculaneum you will see graffiti on the lavatory walls – just the same as if you visit the gents' in Leicester Square. Nearly all graffiti is anonymous, but there is one notable exception to this statement and that is my namesake, Kilroy. James J. Kilroy was employed in a Massachusetts shipyard in the Second World War inspecting warships under construction. To prove to his employers that he was doing his job he wrote KILROY WAS HERE in yellow chalk on everything he inspected. The phrase caught on and began to appear in various unrelated places – and the shipyard workers who joined the forces spread it throughout the world.

Why do people write on walls? I saw the answer in one of the very first bits of graffiti I ever collected, in a London Underground station.

> Why write on a wall?
> *Because it's there.*

If you disagree with graffiti sign a partition!

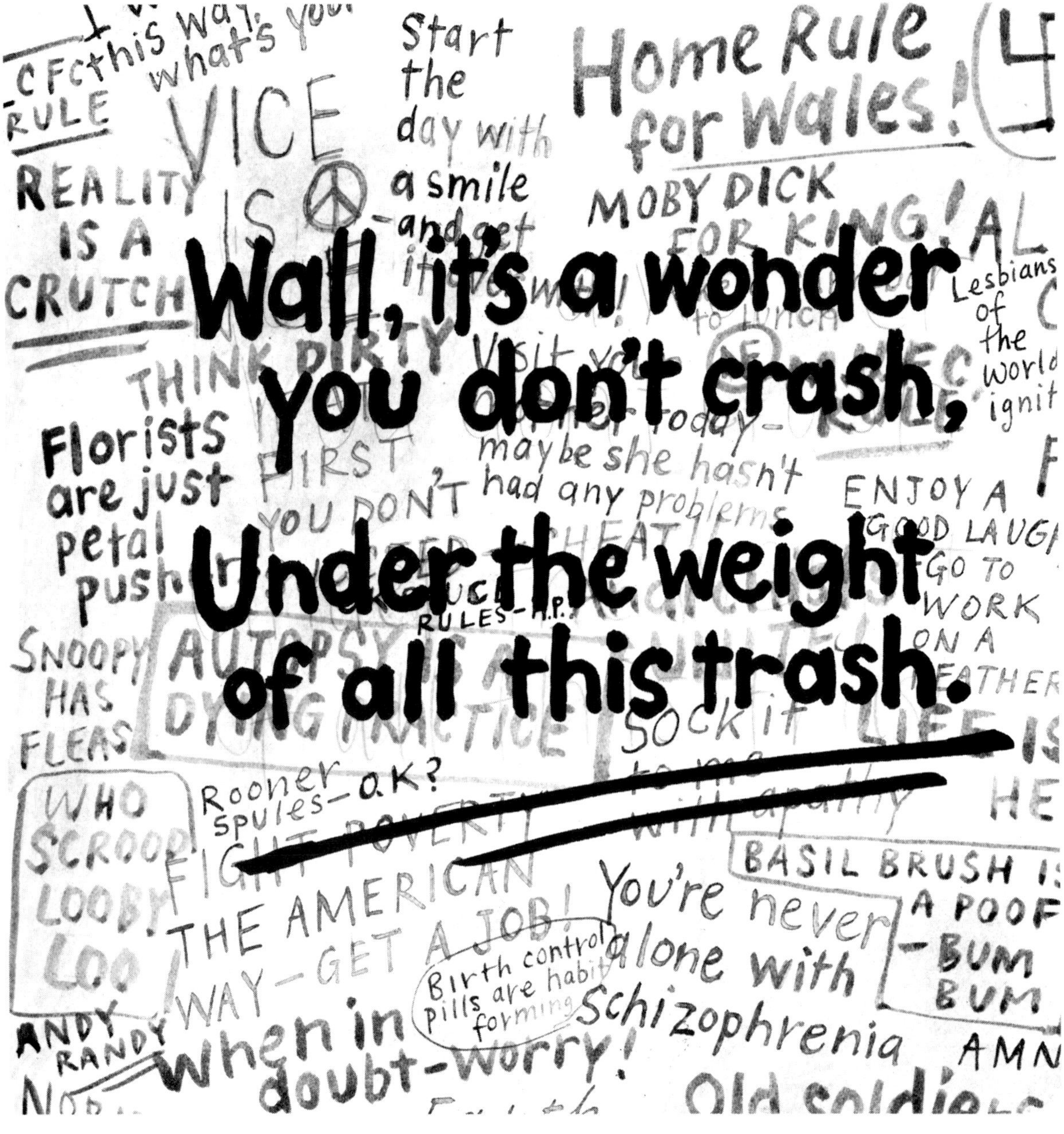
Wall, it's a wonder
you don't crash,
Under the weight
of all this trash.
Start the day with a smile
Home Rule for Wales!
REALITY IS A CRUTCH
MOBY DICK FOR KING!
VICE
Lesbians of the world
Florists are just petal push
maybe she hasn't had any problems
ENJOY A GOOD LAUGH
GO TO WORK ON A
RULES—H.P.
SNOOPY HAS FLEAS
Rooner Spules—ok?
WHO SCROOP LOOBY LOO
THE AMERICAN WAY—GET A JOB!
BASIL BRUSH IS A POOF
—BUM BUM
You're never alone with Schizophrenia
Birth control pills are habit forming
When in doubt—worry!
ANDY RANDY
Old soldiers

I LOVE ~~GRILS~~
GIRLS
I LOVE ~~GRILS~~
GIRLS
What's wrong with us GRILS?

JULIAN SUCKS
ORANGES
I am perfectly well adjusted - Do not fiddle with my knob.
I'm gay and Irish - I go out with girls.
Nowadays Bachelors must insert their masculinity

Don't get mad -get even!!
Laugh, and the world laughs with you
-SNORE, AND YOU SLEEP ALONE!!
Where there's a will - there's a greedy solicitor getting in on the act.
HOW COME IT'S NEVER THE COLD GIRL WHO GETS THE FUR COAT?
ALL THAT GLITTERS IS NOT GOLD
All that doesn't glitter isn't either

ONE
MAN
BAND

Spending a Penny

It was George Jennings, designer of the Gold Medal Award 'Pedestal Vase' who was responsible for this familiar euphemism. He installed public lavatories in the Crystal Palace for the Great Exhibition of 1851 and he felt very strongly that decent public lavatories should be available. To counteract objections on the grounds of cost, he proposed that a small fee should be paid by the user. His public lavatories were very popular, probably because the conveniences that existed previously were communal latrines without seats or partitions – users must have thought the new ones well worth the cost.

Nowadays, too, people are willing to pay for an improved public lavatory. The Superloo, or Automatic Public Convenience as it is correctly named, costs 10p, yet it has proved very popular (average daily usage in 1983 125 people) despite the fact that it is situated near to existing free public lavatories.

The APC is a French innovation, and was first introduced to Britain by Westminster City Council. The first one was opened in Leicester Square on 5 May 1982, and two years later there are twenty in the Westminster area, others in outer London, and they are gradually spreading to the provinces. Westminster Council spends £2 million a year on its public conveniences – mostly on staff wages – and its lavatories are open from 7 am to 10 or 11 pm. It was in an effort to reduce costs by reducing the number of paid attendant hours, as well as to provide a twenty-four-hour service, that they introduced the APC.

The attraction of the APC is that it is completely washed and disinfected between users. The lavatory bowl and floor tip up and are cleaned with a high-speed rotating brush, pressurised water and disinfectant. Hot air dries it, deodorant is spread around the compartment, and the lavatory is ready for the next user. The lavatory door locks automatically during the cleaning cycle, and also unlocks automatically if a user has been in it for seventeen minutes, so there's no possibility of sitting there and reading right through the Sunday papers. A new development is the installing of a GPO line connecting each APC with the company's headquarters so any faults can be directly notified and corrected. It is all a far cry from the pre-Jennings public conveniences.

A cast-iron public lavatory in Bristol, now demolished. The photograph was taken in 1966.

An imposing public lavatory of the early 1900s.

Relief for the ladies, at the Paris Exposition 1889.

Contrasts in style: the ladies' in the Derry & Toms building, Kensington, in 1933

and the gents' in a City club in London in 1907.

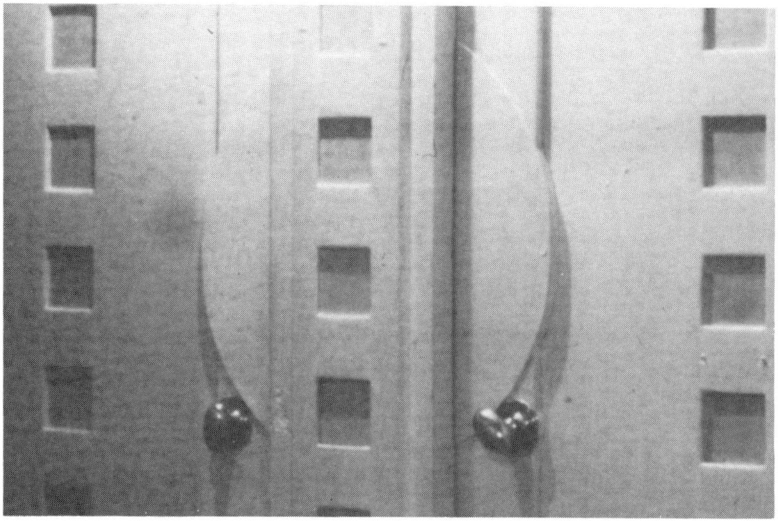

Two lavatory doors designed by Charles Rennie Macintosh: the gents' in Sauciehall Street, Glasgow, early 1900s
and a ladies', also from Glasgow, early 1900s, now in a shoe shop.

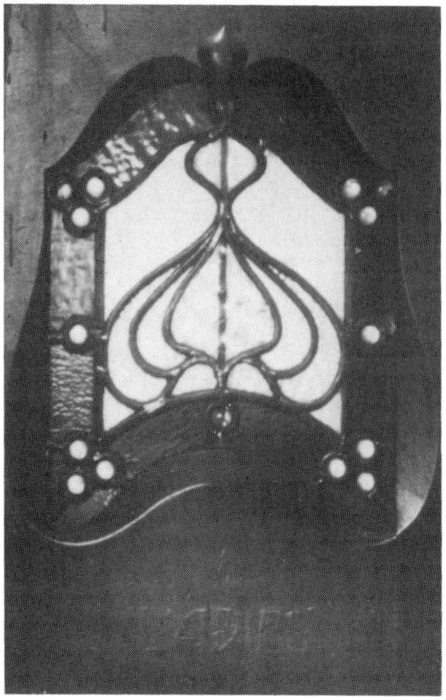

The winners of the first ever Golden Loo Award for the best-kept public convenience in Britain, Reg Bedwell and Les Harding, attendants at the Covent Garden Piazza toilets in London. The competition was organised by Twyfords Bathrooms and the *Sunday Magazine*, whose readers voted for the winning convenience. The attendants personally painted and decorated the convenience, and it is so immaculate that people are said to make special journeys to go and use it.

The Automatic Public Convenience is as near vandal-proof as possible.

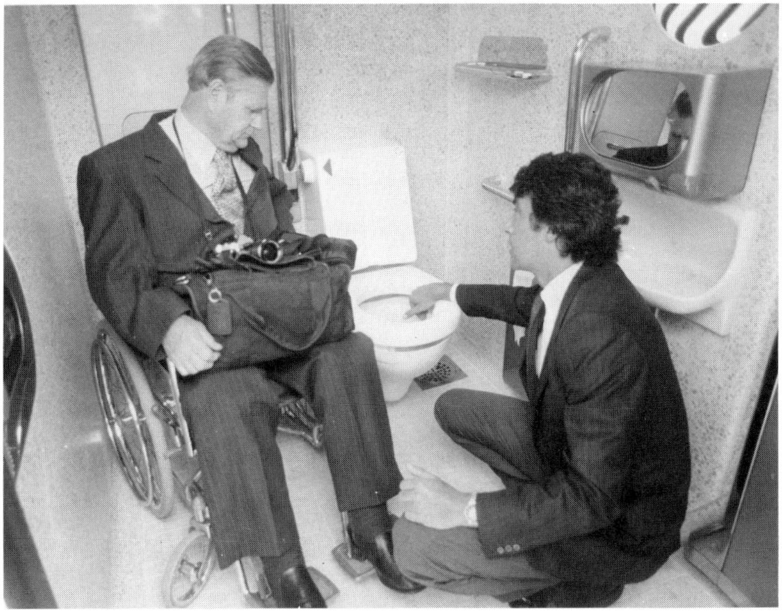

The Automatic Public Convenience being demonstrated to a disabled user.

FOR 1 PERSON

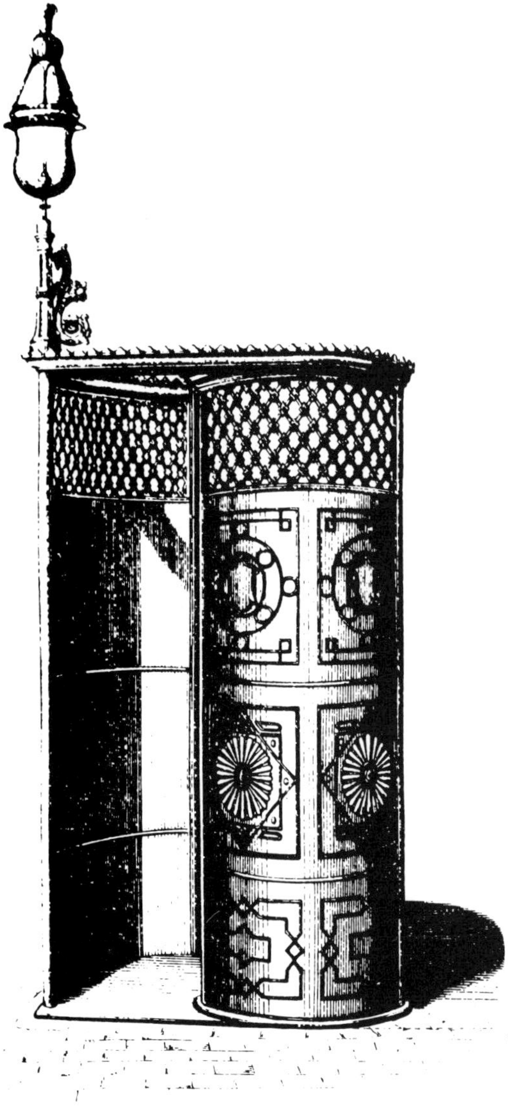

ELEVATION

FOR 2 PERSONS

ELEVATION

Examples of elegant public urinals from a Glasgow iron works catalogue of 1881.

Bottoms Up

A lavatory anecdote by Peter Luke

Time: the beautiful autumn of 1943. Place: the *Campania* of Southern Italy. Naples had fallen to the Allies and Capri had been 'liberated' by Lieutenant Peter Luke and some friends in a rowing boat.

Seeking for further adventure a few days later, Luke was driving along the mainland in a jeep accompanied by Corporal Maggs and Rifleman Bosley when they came to a conical hill dotted about with cypress trees. The top of the hill was crowned by a beautiful villa.

'What about taking a butcher's up there, sir?' said Cpl Maggs.

'Why not?' answered the young lieutenant and, suiting the action to the word, he swung the jeep off the road into a drive that wound round and up the hill in ever diminishing circles. At the top they were rewarded by the sight of an old but luxurious villa set in a garden with a fine view of the sea.

The house seemed untouched by the destructive hand of war and, to their even greater surprise, they found the door unlocked.

After many months spent in the deserts of North Africa these Eighth Army men, on arrival in Italy, had been overwhelmed by the superabundance of autumnal fruits: everything from grapes and pear-shaped yellow tomatoes to persimmons and ripe walnuts, and their digestive systems had not yet become accustomed to such bounty. Accordingly, on arrival at the villa, Peter Luke, who was feeling the call of nature, straightaway went in search of a lavatory. He soon found one – a beauty, bright with ceramic tiles and garnished with colognes and toilet waters, sea-sponges and every sort of bathroom accessory.

Luke was bending over on the point of sitting down when the loud crack of a rifle-shot instinctively caused him to pull his pants up again. Outside he came upon Cpl Maggs easing a spent round out of the chamber of his Lee Enfield.

'Missed the bugger! Two Jerries – eh, Boz?' He turned to the rifleman beside him for confirmation.

'Yeah, Jerries. Two of 'em,' agreed Bosley. 'I thought there wasn't supposed to be none south of the Volturno.'

'Deserters, maybe,' said Cpl Maggs.

'I don't care who they were. They've spoilt my crap in one of the most beautiful shit-houses I've ever seen.'

'Go on, Guv'. Let's 'ave a shufti, then. I've forgotten what a bathroom looks like.'

The three men went back into the house and looked around appreciatively.

'It's a bleedin' palazzo, that's what it is,' asserted Bosley authoritatively.

'Let's 'ave a decko at the carsey, then,' said Maggs. Luke led the way into the bathroom.

'Marble floors an' all,' said Bosley.

Suddenly Maggs's attention became focused on the lavatory.

''Ere, wait a minute,' he said, crossing over to it. ''Old on.'

After a brief scrutiny, Maggs gingerly lifted up the seat to reveal a piece of mechanism taped to the pan. Following a wire attached to the mechanism he found it led to a box concealed behind the bowl.

''Old everything,' said Maggs, and slowly prized open the lid of the box. Inside it was one of the most destructive pieces of military hardware for its size at that time known to man – a German S-mine.

'My God!' said Luke, inadvertently putting his hand to his backside.

'Yeah,' said Maggs, 'and he'd be the first bloke you'd be seein' if you'd sat down on that seat.'

Carefully Cpl Maggs disconnected the trip-mechanism.

'There you go, sir. You can enjoy yourself now,' he said.

But Peter Luke was already half-way out of the bathroom.

'Never again', he said. 'I shall never be able to sit on one ever again.'

And he walked out into the soft autumn sunlight and disappeared into the bushes.

TWO LOOS
SOUTHERN FRANCE
MESSIEURS
Café

'Some Corner of a Foreign Field . . .'

The English are very scathing about foreign plumbing, doubtless not realising that until comparatively recent times they had nothing to be proud of in their own country. Continental loos are enough to discomfort the average English traveller – how much worse, then, those of the undeveloped countries? The proper disposal of sewage is one of the most pressing of public health problems – hundreds of millions of people are affected by inadequate sanitation and suffer ill-health, poverty and death as a result. Because of this the United Nations declared 1981–90 the International Drinking Water Supply and Sanitation Decade, and Oxfam is installing sanitation units consisting of twenty latrines in two rows of ten separated by a partition capable of dealing with the needs of 1000 people a day.

For those who are unacquainted with Third World sanitation, here are two very different ways with which they cope with the problem.

First, an account of a ventilated pit latrine which was invented by Dr Peter Morgan of the Blair Research Laboratory, Ministry of Health, Zimbabwe.

The Zimbabwe Long Drop

by Louise Westwater

The Zimbabwe Long Drop is a pit latrine that is fly-free and odour-free. It is odour-free because of the draught created in the ventilation pipe as it heats up in the sun, allowing the air to go in the door, down the hole and up the ventilation pipe and not the other way round. And it is fly-free because flies are attracted by *light* and *smell*, both of which are minimised. The smells come out the top of the ventilation pipe, which is covered by gauze so flies cannot get in or out. Any flies that do get into the pit are not attracted to come out of the hole, which is *dark* but go up the ventilation pipe, where they are trapped by the gauze and quickly die in the extreme temperature.

The pit should be 15 ft deep and 6 ft across. Experience has shown that a pit of this size will never fill up if used by one large family. Over the pit is placed a concrete slab, which is cast on site. The slab has two holes: one for the ventilation pipe, on the edge, and one in the middle for the user.

The size of the squat hole is critical. It must be large enough so that an adult can use it without fouling the edges, but not so large that it frightens the children. African children from the age of about two and a half use a toilet. But if they see a huge black hole that they cannot straddle easily, and may fall down, they won't use it and instead will use a corner inside the latrine or go outside round the back of it.

The inside of the toilet must be dark to avoid attracting flies.

The entrance does not have a door. Doors need hinges. Hinges get broken, doors fall off and are used as firewood, the toilets are no longer dark, nor private, and so are no longer used. Instead the entrance is in the form of a spiral. There are some big ladies in Africa, so the spiral and the space in the middle must be big

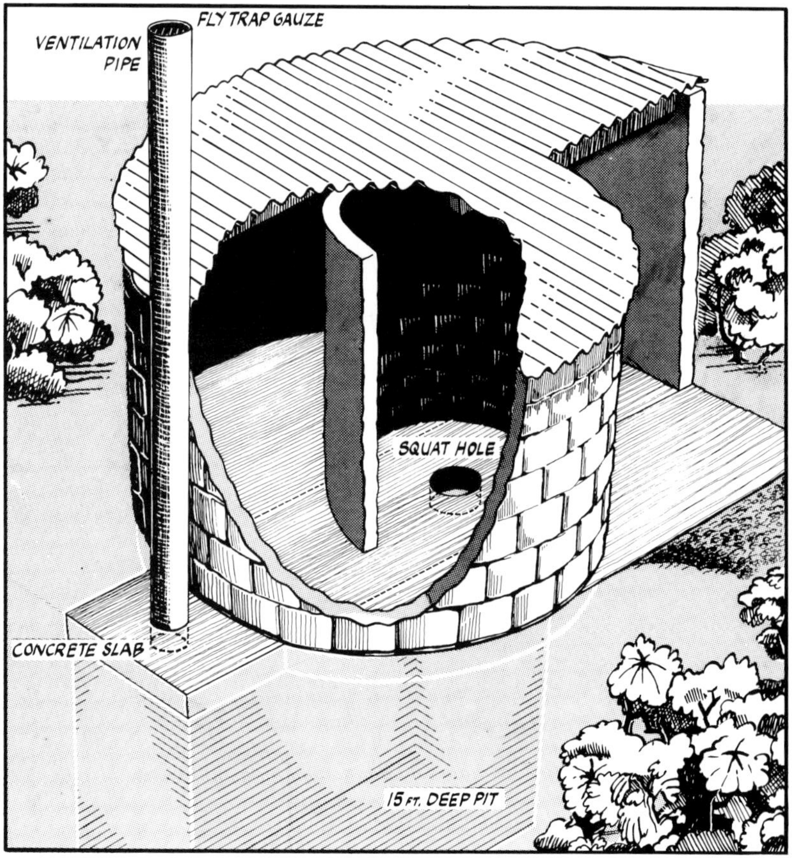

The Zimbabwe Long Drop.

enough for the largest lady in the family to go in, turn round and come out again without getting stuck.

The superstructure must be of a permanent material, not 'pole and dagga' (wattle and daub) and thatch. Home-made brick is suitable. The roof must be manufactured from outside materials. Asbestos cement, despite its bad reputation in Europe, is best and cheapest. The Zimbabwe asbestos from Shabani is long-fibred and not carcinogenic.

The size of the ventilation pipe is critical. Misguided efforts to economise using a small-bore pipe render the whole system ineffective. The draught is not created, the air current does not flow and the smell then is just as bad as from any other pit latrine. The pipe must be 9 in in diameter, and placed on the outside of the building on the sunny side. In Zimbabwe the sun shines from 6 am to 6 pm without fail every day, winter and summer, so the system always works, and the temperature of the air inside the pipe is very high indeed. A nice demonstration using a smoking rag shows the smoke sucked in through the spiral, down the hole and within seconds appearing out of the top of the ventilation pipe.

This simple system was developed by a clever young man, trained originally in social science, which may be why he managed to produce an apparatus which does not require any modification of the users' behaviour. Too many devices for use in Africa are ill-thought-out – they require modification of human behaviour in order to work, they have breakable parts, they need running water, or they can be used only by adults, for example, the domestic tap and the flush lavatory. Many toilets have been a disaster in Africa because the European inventor forgot that 25 per cent of the users are small children with short legs that can't straddle a big hole, and short arms that can't reach the chain. The same European also forgot, or never knew, that no self-respecting rural African will sit on a lavatory seat, so if you provide a pedestal with a seat he will squat on that. The Zimbabwe Long Drop can be constructed in a rural area without running water, and from materials costing between £30 and £50, most of which is usually available from Aid, government, or the local co-operative, if the labour is provided by the family.

Second, a description of lavatories on the island of Badu, in the Torres Strait between Australia and New Guinea, by Lucy Irvine, author of *Castaway*, a book which describes the year she spent on nearby Tuin island.

Lavatories on Badu
by Lucy Irvine

Lavatories, like God and the common cold, arrived on Badu with 'the coming of the light' in 1871. In those days they were picturesque, palm-thatched structures covering holes and placed at a discreet distance from the main settlement. But this did not suit the gregarious nature of the islanders and soon they moved the 'pee-pee ouse' to a prominent position and raked a large area of sand in front of it clean. When, early in the morning after a dugong feast, the chief's Numbawan wife whispered to the Numbawan son who in turn whispered to the Numbawan messenger who whispered to the nearest running boy who ran up a palm tree and blew a conch – there would be a prompt but respectful convergence on the pee-pee ouse. The chief was about to perform.

At first, these gatherings were merely social, but at least one headman was quick to realise the advantages of such a position and used it to preside over meetings of the elders – rather unfairly forcing other members of the community back to the bush and a banana leaf as not everyone could wait until he had unburdened himself of all the weighty matters of the day.

These days, with the advent of chipboard, corrugated iron and white man's manners, some of the original attraction of the pee-pee ouse has been lost. But old habits die hard and although the Australian government thoughtfully provides doors for each one of the grey, malodorous sentry boxes now dotting the island, these are usually torn off to give a better view, coming in handy as duckboards during the wet season. Unfortunately, they are also sometimes used to scrape the worst off small children who fall into the black, rubbery cans which, to accommodate the big-is-beautiful cult of the adults, have perilously wide mouths. Toilet paper is not provided except around mainland election time when political pamphlets are dropped in by helicopter.

The position of 's'it fella' is considered a Numbawan job on

Badu as with it goes the privilege of driving the council tractor. All day long he bounces blithely over the unmade roads enveloped in a torrid miasma of heat, dust and bloated flies. A permanent gang of joyriding ladies keeps him company. Every now and then he stops to check the level of maggots in one of the communal pee-pee ouses and prods out any goanna or carpet python foolish enough to have ventured in. When the yellow, heaving mass reaches within an inch of the top of a can, he claps on a lid and swings it on to his trailer, where it is immediately pounced upon by one of the ladies. A big cheer goes up when all fourteen have a throne.

At the end of the day the s'it fella trundles his load down to the sea and the sun sets on a vision of the ladies bathing while further down the current Badu's Numbatwo floats off to Papua New Guinea.

'An' that why,' I heard one present-day elder remark with a gleeful snigger, 'them fella got skin more black than we!'

A 'pee-pee ouse' on Badu.

A health education poster displayed in public conveniences in the Solomon Islands.

A typical French *pissoir* of the old style – in the town square in Doullens.

Is this the world's tallest loo? It is situated in an abandoned brick factory chimney outside Grahamstown, South Africa, and was provided for workers loading bricks on to trains near the old brickyard. The workers commented: 'It gives us an excuse to meditate on old times.'

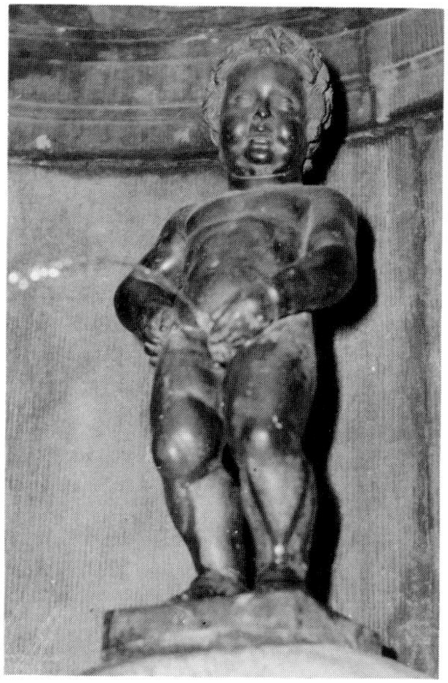

The *Manneken-pis* statue in Brussels, which seems to sum up the Continentals' attitude to the whole subject.

Inspired by the *Manneken-pis*, a cartoon of Margaret Thatcher by M. Hoviz of Paris, after the failure of the EEC talks in Brussels.

RODIN'S
CHANNEL
FERRY
TOILET

In Motion

Ever since man began to travel – other than on foot or by horse – he has needed some kind of facility for relieving himself *en route*; if not a chamber pot in his carriage then an accessible public lavatory, or comfort station, as the Americans say. Nowadays even coaches have their own lavatories, but no one seems to have yet devised one for use in a car.

Most of us are familiar with the train loo, with its unpunctuated motto 'Gentlemen lift the seat'. Train loos are not very inviting places, but those on grand trains like the Orient Express were splendid affairs, with mosaic floors depicting scenes from Greek and Roman mythology.

Lavatories on ships used to be called 'heads' because they were situated at the front or head of the ship. Here they were exposed to the wind and would be washed by the waves if the sea was rough.

Going to the loo on a plane is a fairly uncomfortable experience – but what do people do in space? Astronauts have to cope with the problem of weightlessness – if there is no gravity both liquids and solids simply float about. For this reason the men used to have to manage with a kind of plastic bag arrangement inside their space suits, but nowadays they can take the suits off and use vacuum toilets which flush by air instead of water and suck away the waste as it leaves the body. Liquids evaporate and solids are freeze-dried, sealed in packs and returned to earth. It's a relief to know that, isn't it?

A cast-iron gents' on a platform at Bewdley Station, Severn Valley Railway.

Mosaic floors of the loos on the Orient Express depicting characters from Greek and Roman mythology.

The 'heads' on the deck of the *Victory* at Portsmouth.

The 'heads' on a submarine, HMS *Thermopylae*, launched 1945, and their operating instructions. This apparatus is now in the Royal Navy Submarine Museum, Gosport.

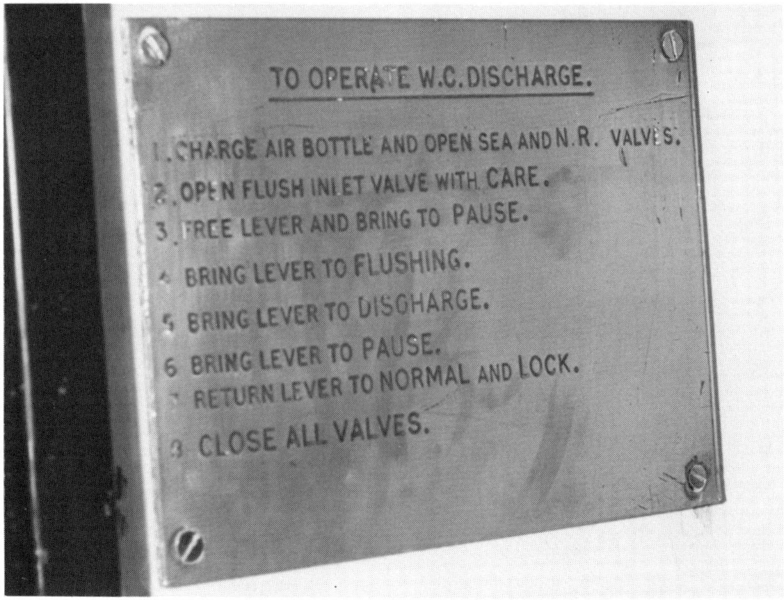

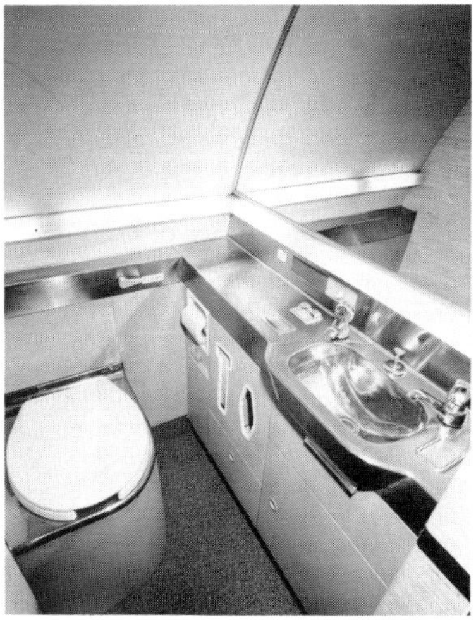

Padded comfort in the air on a 1-11 Lanica, left; and the facilities on a 1-11 Air Malawi plane.

Seats to Suit or what some people's loos *ought* to look like.

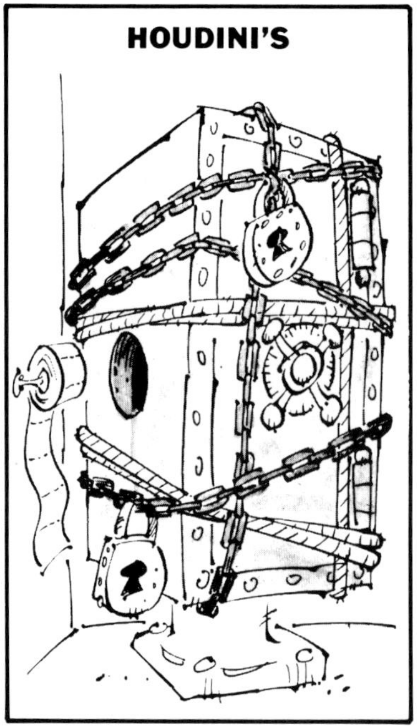

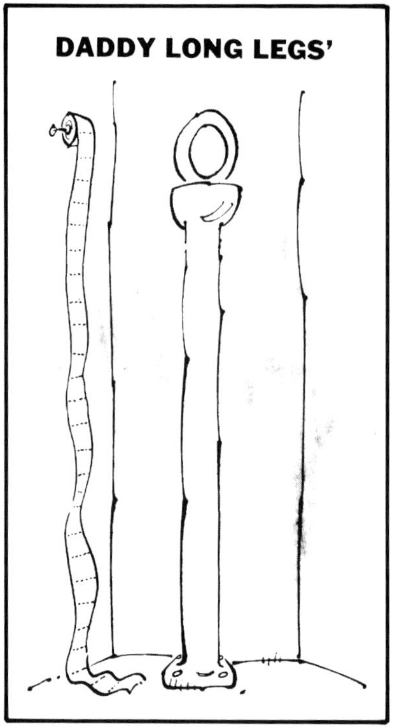

from Mad Magazine

DOULTON
ROVER

A Lamp-post By Any Other Name

From time to time local papers in big cities receive batches of indignant letters about the state of public parks, thanks to dogs and their inconsiderate owners, and battle rages between dog-owners and parents of young children. To try to combat the menace, various councils have experimented with dog loos. In some cases these have been whimsical affairs, with a lamp-post standing in a sand pit, but mostly they are just fenced-off sand pits, with notices proclaiming their function. They are cleaned out by the authorities once a week. Unfortunately they have not proved to be very successful, with well under 50 per cent of dogs using them, though conscientious owners do try to train their dogs to do so. Some councils have now abandoned the experiment.

The state of our cities' pavements also causes concern, and a walk round the block can be said to be a time of seeing all the old familiar faeces, though those who complain about dogs often seem oblivious to the greater mess made by human beings. But in the nineteenth century collecting dogs' dung was a profitable business. Collectors were called 'pure finders', and they sold the dung to tanners, who rubbed it into animal skins to help 'purify' them. This was probably one of the reasons why tanning was considered to be an occupation pretty low down on the social scale. However, New Yorkers with their 'pooper scoopers' must sometimes wish that they could obtain some benefit other than a feeling of righteousness from performing their noisome task.

Most animals in their natural state reserve a special place for their loo, and domesticated animals often follow suit, for example, horses often soil only one area of the stable. All cat owners will be familiar with their pets using a particular place for the loo, and make use of this habit to train them to use a litter tray when confined to the house. Nowadays dog owners too can contain the mess their pets make at home. They can buy a dog loo. This is not a device that the animal actually squats over, but a receptacle for the waste which is chemically dissolved into a harmless liquid that filters out into the ground. A pit is dug to accommodate the loo, the chemical is added, and from time to time the whole thing is flushed out with water. Despite the fact that the owner still has to shovel up the mess, it is probably the best solution that exists for dealing with an unpleasant problem.

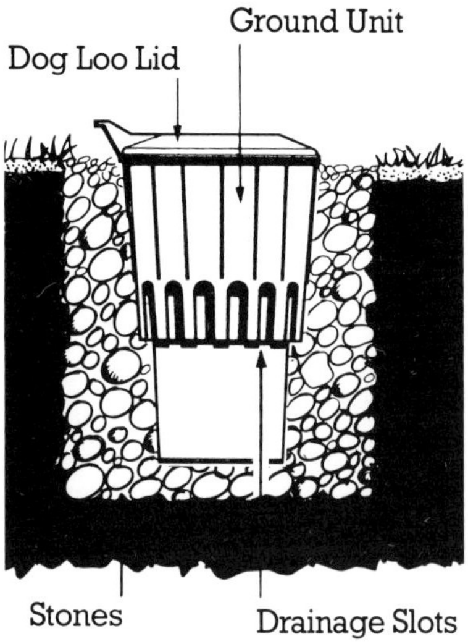

A diagram of the dog loo.

A dog loo in Marble Hill Park,
Twickenham, Middlesex.

On Paper

Lavatory paper, or as it is now labelled, toilet tissue (240 sheets per roll) is something on which the British have very definite views. Although in their homes they may use soft paper, in public they stick to the hard stuff, either on economic grounds, or because it is the sort of comfort – like central heating – that the British stiff upper lip views with disfavour. All establishments maintained by the state have 'Government Property' stamped firmly across their hard loo paper, and it is only in the higher echelons of the Civil Service that toilets are provided with soft paper.

Considering the possibilities it is surprising that more people have not produced loo paper embellished with scatalogical verse or drawings, though around Christmas time one does sometimes see advertisements for paper bearing the face of, for example, Margaret Thatcher.

People have used many things for loo paper over the years, from the medieval monks who appear to have used scraps of old gowns to the Muslims who never let their right hands know what their left hands do; from dried leaves and corn cobs to the stones used by some Africans; from the sponge sticks of the ancient Romans to the sheets of newspaper and other reading matter used, at times, by practically everybody.

Probably the best-selling book of all time about lavatories, *The Specialist* by Charles Sale, described using thick mail-order catalogues, a new one in January lasting until harvest-time, with luck and not too many visitors.

This little book was first published in 1930 and is still in print. It has sold over one million copies in the USA, where it was first published, and over 700,000 in Great Britain. It is published by Putnam & Co Ltd.

Another best-seller, brought to the attention of the British public by a television series, is *Clochemerle*. Published by Secker & Warburg, it sold 1500 copies in its first month after publication in 1936, which was considered excellent for a first novel, and it is still in print and selling steadily. It must prove something about the fascination exerted by the subject on people's minds.

Some examples of toilet roll packaging.

The jacket of *The Specialist*, a best-seller about the professional activities of Lem Putt, a privy builder with an unusually good understanding of human nature.

GABRIEL CHEVALLIER

Clochemerle

LES ÉDITIONS RIEDER - PARIS - MCMXXXIV

The title page of the French edition of *Clochemerle*, a best-selling saga about the siting of a French public lavatory.

NATIONAL EXHIBITION CENTRE
LOOS OF THE FUTURE
CAN'YOU TELL ME WHERE THE NEAREST GENTS IS?

Future Perfect?

And what of the future? Loo design has not changed a great deal in recent years, sanitary manufacturers concentrating instead on producing glamorous baths. But there have been some developments.

A Swedish manufacturer has produced a loo that can flush with as little as one litre (two pints) of water, compared with the average British flush of two to three gallons, though it is being marketed in the UK in a 6-litre version. It has a specially shaped bowl to enable it to empty quickly.

The invention of a small-bore outlet system allows loos to be installed where there is no room for the conventional 4-in discharge pipe. A device fitted to the rear of the loo shreds the waste and paper to a slurry which is pumped out through a ¾-in waste pipe; the pumping action means that it can be installed in basements and other places where the waste water would otherwise have to flow uphill.

Some of the developments are installed in private homes on the Continent but in England used mainly as an aid to the disabled. This is partly on the grounds of cost, and partly because the British water bye-laws are more stringent than their Continental counterparts. One such development is the Swiss automatic loo, which combines a loo and bidet in one apparatus, washing the user with warm water and drying him or her with warm air, thus doing away with the need for loo paper.

Another Swiss invention was launched in 1980 by engineers who wanted to produce the most hygienic loo seat possible. It consists of a polythene sleeve which encases the seat and is wound round it from spools on either side, rather like the film in a camera.

And finally, there is news that Thomas Crapper's Valveless Water Waste Preventer may have almost had its day. An inventor called Glyn Roberts has produced a device called the Ve Cone valve, which, its manufacturers proudly proclaim, could have the same effect on plumbing as the silicon chip has had on electronics. The search for perfection goes on.

The Saniflo small-bore system by Transbyn, which enables a loo to be installed in places where a conventional waste pipe cannot be fitted.

The Clos o mat 'Samoa' automatic WC which combines the functions of a loo and a bidet, pressure on the levers activating flushing, washing and drying with warm air.

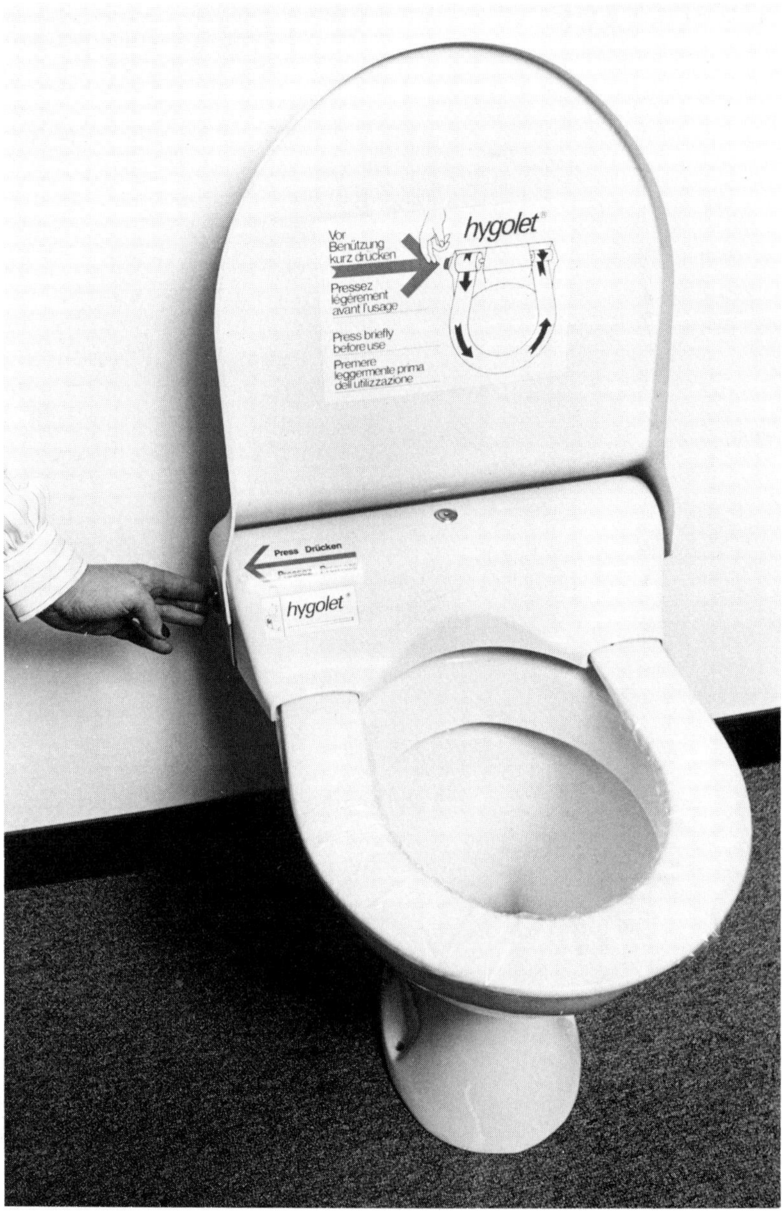

The Hygolet system for producing the most hygienic loo seat possible.